HOP
PICKERS

Please return on or before the latest date above.
You can renew online at *www.kent.gov.uk/libs*
or by telephone 08458 247 200

CUSTOMER SERVICE EXCELLENCE

Libraries & Archives

00884\DTP\RN\07.07   LIB 7

HOP PICKERS

of

KENT & SUSSEX

HILARY HEFFERNAN

First published 2008

The History Press Ltd
The Mill, Brimscombe Port
Stroud, Gloucestershire, GL5 2QG
www.thehistorypress.co.uk

Reprinted 2008

British Library Cataloguing in Publication Data.
A catalogue record for this book is available from the British Library.

ISBN 978 0 7524 4777 3

Typesetting and origination by The History Press Ltd.
Printed in Great Britain

Contents

'Autumn after hop picking' – a watercolour painting of a hop farm by the author.

DEDICATION AND ACKNOWLEDGEMENTS

This book, my fifth about hops and life in the hop gardens, is dedicated to my dear grandchildren, Dominic, Holly, Damasca and Zane, with my love.

My sincere thanks go to all those hoppers, friends, brewers, hopping farmers and others who shared their reminiscences as well as coffee and cakes with me from those happy days down on the hop farms, particularly my dear friend of thirty-five years, Sidney Fagan (who has long encouraged me to put pen to paper and was himself a hopper as a child). Also, Ronald and June Sparrow, Terry and June Hogger, Marion Healey, Carol Goodman, Steve Knutt (who suggested the idea of an audio book; an idea I will be taking up later this year so watch out for it!), Ben and Marion Healey, Alice Heskitt and all my other contributors.

I have enjoyed writing their stories and hope that, as you read it, this book recalls many happy situations from your own lives down hopping.

I apologise to any person or company whose contribution I have included and inadvertently omitted to acknowledge.

Hilary Heffernan
June 2008

HOPPING DAYS

They still stand tall,
Those chestnut poles,
Established in their tar-filled holes.
But when in August or September
I pause a while, just to remember,
No neat cones of palest green
On twining bines can now be seen.
The air has lost the scent of hops
As bines fall when the puller lops.
The chill of that early morning air
And dewdrop diamonds in my hair
As we kids made an early rise
At the foreman's cheery 'All out!' cries.
And as the daylight hour expires
We gather round the evening fires
And sit on bales and the hopping cart,
Then sing, and sing with a happy heart.
Now the green has swiftly turned to grey
And hopping, sadly, has had its day.
But memories live here in my mind
And happier days we'll never find.

Hilary Heffernan

INTRODUCTION

HAPPILY HOPPING DOWN IN KENT

Kent, Sir? Everybody knows Kent! Apples, cherries, hops and women.

Quote by Mr Jingle in Dickens' *Pickwick Papers*

The humble hop cone once offered thousands of underprivileged Londoners the annual chance to leave London's smog in August and September, migrate to the healthier countryside and live happily for six or seven weeks in near squalor, sleeping on straw-filled palliasses crawling with mice, earwigs and spiders and cook their dinners on open fires – even in the pouring rain. All this while arising in the early mornings to arrive dressed in wellies and their oldest clothes at muddy, bedewed hop gardens to pick hops for as little as 5d per bushel (a bushel basket has the capacity of a large waste-paper bin) and generally have the time of their lives. Despite the acknowledged hardships, talk to an ex-hop picker and you will nearly always find a wealth of happy memories and someone who was content with their lot. This is not everyone's dream of pleasure, but at one time many thousands of workers looked forward to the annual exodus from London and nearby towns to go down to the hop gardens of Kent and Sussex. It was the highlight of every year. 'We couldn't wait', says Mrs Sarah George, who now resides at Crystal Palace but lived at Bermondsey as a child:

We kids started getting excited from the moment that letter arrived from the farmer to say he wanted us down there the third week in August and we spent from the time it arrived until it was finally time to start filling up the 'opping box and all the excitement of planning what we'd do once we got down there at the gardens that season. Our mum, grandma and aunts did nearly all the picking but after we little ones had picked for a couple of hours before they'd say 'Off you go, and keep out of mischief, we don't want the farmer after us!', and we were free to explore the hedges for hens eggs, climb trees and go down to the duck pond to fish with a stick and a bit of string tied on with a bent

Above left: Bushel baskets.

Above right: Hop pickers arriving at the hop farm.

pin dangling off the end, or dip for pond skaters in the water with a jam jar with string tied round the neck for a handle until mum called us up for dinner.

Within living memory, vast areas of Kent and Sussex were covered with either hop gardens (also known as fields and yards) or orchards. Hops, once prolific across Kent, Sussex, Worcestershire, Herefordshire and a few other English counties, are now a relatively scarce commodity in what is still known as 'the Garden of England'. Hops tend to come in pelletized form as exports from the continent. Nowadays, the 25ft-long bines, straggling forlornly over hedges as a wayside plant, are not always recognised by the casual rambler as their basis for a tasty pint of beer. They once flourished by the acre in neat, serried rows, strongly supported by purpose-planted chestnut poles with wires stretching along their tops to support the weight of heavy bines. A thousand voices rang out down country lanes as hoppers joined in a Chorus of the latest songs or whistled a cheerful accompanying refrain. Come to think of it, you don't often hear people singing or whistling as they work nowadays!

Whitbread's Hop Farm, now the Hop Farm Country Park at Paddock Wood, once employed as many as 4,000 pickers at a time for their seasons. Would-be hoppers poured onto trains departing from London Bridge Station at midnight the night before picking was due to start, steaming their way down through Kent and Sussex in the dark to arrive in the early morning at all stops where hop farms could be found along their various routes. For the children it was a real adventure, not second-hand thrills and painless spills via a computer or Game Boy, which is all most of today's children have. For the adults, it was an opportunity to earn a bit of extra cash and break free from a hard life in the City. The farmers made ready for this vast annual influx by providing hop huts or barns of varying qualities – wood, tin or brick – for the temporary use of

Right: Geese were a novelty to town children.

Below: Rules of the hop

WHITBREAD & CO., LTD.,

BELTRING, PADDOCK WOOD, KENT.

HOP-PICKING.

RULES AND REGULATIONS TO BE OBSERVED BY BINMEN.

1. Binmen must be in the Hop Gardens at 7 a.m.

2. They must provide a constant supply of bines for six bins. All bines must be cut by the Binmen at the top wires, and none should on any account be pulled down, either by the Binmen or the Pickers. All tops, or branches, left on the wires after the bine has been cut down must be taken down at once, and not allowed to hang and wither.

3. Binmen are responsible for seeing that the ground is cleared of dropped hops, and pickers should not be allowed to trample hops underfoot.

4 Binmen must look after their pickers, and see that they pick the hops cleanly and free from bunches. All hops must be picked from the bines, unless the Hop-garden Manager orders otherwise. When finished with, the picked bines should be wound neatly round the hill stumps.

5. Binmen must carry out their full pokes and load them on to the wagons immediately after each measuring. They must not put them down and leave them. No child or boy is to be allowed to fetch or carry pokes, either empty or full.

6. Binmen must be particularly careful to see that 10 bushels of hops, neither more nor less, are put into each poke. They should see that the pokes for the night loadings are tied loosely.

7. When taking up fresh ground, a Binman should see that all the members of his Company move together.

8. Each Binman is responsible for his pokes, knife, lapbag and bincloths.

9 No Binman may on any account take the pokes of another, without first having obtained permission from the Hop-garden Manager, or his Assistants.

10. Binmen must immediately report anyone breaking the rules, stealing or doing wilful damage in any part of the farm.

11. No Binman may leave the Hop Gardens during working hours, without special permission.

12 Binmen's wages are fixed at............per day for each completed working day, which may, if desired. be subbed on the usual days. Binmen are engaged by the day, and may be dismissed at any time without notice. An additional sum of 6d, per working day will be paid to each Binman who is not discharged for unsatisfactory conduct, and remains throughout the picking.

Special Notice.

In the past we have found some Binmen have been inclined to neglect their proper work to pick hops. We wish all Binmen clearly to understand that they are allowed to pick as a favour, and only on the strict understanding that such picking does not interfere with any of their proper duties.

families, some of whom regularly used the same hut year after year. At some farms the farmer had to turf out the animals that occupied the huts, such as pigs, hens or horses, and give them a quick swill down before the hoppers arrived and finished cleaning them properly before they could occupy them. Whitbread's brick huts, with electric light laid on, were considered the height of luxury compared with the corrugated-iron huts, barn space or wooden shacks provided on most other farms, but these were demolished in the 1980s when the Medway River flooded, devastating much of the surrounding countryside late in the twentieth century. At least one farmer used his pig huts for hoppers, which meant turning the pigs out into his fields shortly before the hoppers were due, hosing down the interiors of the huts and leaving his newly arrived hopping workforce to do the rest for themselves when they reached the farm.

A few hop huts still survive on former hop gardens. In some cases they are let out to families who were former pickers on a particular farm and are now used by them as holiday homes. The majority, sadly, are in a dilapidated condition or have been left to fall into total decay over the years and are no longer habitable. Other hoppers slept in barns, making a private place for their family by slinging blankets on ropes across the width of the internal space and sleeping on the straw or hay. It was common to see hoppers arriving at a farm in early August, armed with pots of whitewash or distemper and brushes, ends of wallpaper rolls or sheets of newspaper to paste on the walls and handmade shelves, ready to transform the inside walls of their huts. Some preferred to use small, decorative pages taken from out-of-date wallpaper sample books begged from shopkeepers, or illustrated pages from picture magazines. The paste to stick them on the walls was made from flour and water. My mum and dad, Dot and Walter, made a very serviceable rag rug in the 1940s. They got together a heap of old clothes, cut the material into strips and, using an opened-up sack as a base, folded each strip in half and hooked it twice through a hole in the sack, then knotted the strands by the simple process of slipping the two strip ends through its loop and pulling it tight. It was dyed black, and a pattern was added later as a decoration by adding a shield shape of red-dyed cloth strips. Sometimes lace curtains were hung at doors or windows for privacy, as this allowed the occupants to leave their hut door open on hot summer days. Whatever their ideas of interior design, the hoppers soon had their huts looking cheerful, comfortable and ready for occupancy by the time the actual business of hop picking was ready to begin. They had to complete it soon after arrival, as once the business of picking began, there would be little time for anything else. Farmers seldom interfered with interior decoration of the huts. As long as the hoppers did good work they were content to leave the workers to make their hut their home, even allowing them to leave bits of furniture and gadgets there, such as a Valour paraffin stove, from year to year. Huts didn't have locks on the doors – there was no need to keep thieves out as there was nothing valuable to steal.

Walks in the countryside have always been popular, and ramblers could easily tell when they were approaching a hop garden due to the powerful smell of hops. As well as permeating the clothes, hands and hair of pickers, the distinctive, clean smell of hops gave a strong herbal odour to the countryside. Part-way through the year they gave off an aroma of nicotine when crops were sprayed against fungal infection. If any of the plants developed a fungus the whole crop had to be destroyed so as not to spread the disease to other gardens, so spraying was an essential ritual of growing hops. Because of this threat, hop farmers needed to have a contingency against crop failure and this usually took the form of fruit growing, pigs, dairy cattle or root-vegetable crops that they could fall back on if times were hard. This practice caused its own problems, as orchards and vegetables were usually planted in fields next to hops, which was handy for the ever-hungry pickers who surreptitiously supplemented their meagre meals with fresh fruit and vegetables, which they doubtless could ill-afford in the city. At least one farmer even went so far to protect his crop as to invent a convenient ghost purported to haunt his orchards, hoping to scare off determined scrumpers.

Sympathetic farmers thoughtfully supplied their hoppers with a wagonload of windfalls from the orchards and told them to help themselves; some looked the other way when workers sneaked into their crop fields late at night or in the early morning before the sun rose and filled pockets, a bowl, bag or whatever was available, with fresh apples, carrots, cherries or turnips and sneaked them back to their huts as the chilly morning mists curled steamily above the hedgerows. Others were not so forgiving and if a family was caught scrumping (stealing fruit or vegetables) while on the property of a strict farmer there could be severe penalties. They may just be given a warning the first time, but the second transgression usually meant dismissal in shame from the farm. This action not only affected the whole workforce, as it brought home to them that scrumping may be good fun for the children and useful for supplementing the larder, but it was a strong warning as to what would happen to other scrumpers if they were caught what was, in fact, stealing. Any dismissal had a devastating effect on the whole family, as not only were they disgraced as thieves before all the other hoppers, but they were sent home at their own expense, lost their annual holiday, were not allowed back to the farm the following year and the loss of extra earnings was a dire blow to the family's overall annual income. It was a prime case of warning by example. If it became known that a particular family were inveterate scrumpers or troublemakers, then the warning was passed around between other farmers and the culprits could find themselves barred from farms in that area. This meant they may have to travel as far as another county to get seasonal work, as well as losing their particular hut.

Huts were valued according to their position on the farm, the length of years a particular family had gone hopping at that farm and whether or not the farmer was in the habit of allowing a family to use the same hut year after year and keep their possessions there, such as a cupboard, cooking pots and other essentials. A prized hut was one not too far from the site's one water tap, or facing east for the warmth of the morning sun, or maybe because it was near a friend's hut. It was not unusual to find that Mrs Brown, Mrs Green and Mrs Smith lived next to each other in their terraced houses in Bermondsey, and that their huts were next to each other when they went hop picking. They would have used those same huts through several generations. Not only were the farmers happy that they were employing workers they knew to be reliable pickers but the pickers themselves were content to go back to an employer whose ways and family they knew, in an area that had become a second home to them over the years. Often they were invited back as friends and it was common for a warm camaraderie to develop between farmer and regular employees.

Families which regularly picked at the same farm each year and became known as reliable workers may also be asked to stay on for potato, strawberry or apple picking after the hops were gathered in. This was useful as it could mean a further two weeks' paid 'holiday'. It was mainly women employed in this work as their husbands and older sons and daughters were required to attend their regular schools and jobs. A week's holiday was the most paid annual leave a man could expect from his place of employment until the 1960s, when employers became more liberal in allocating holidays to their workers. To take further time off meant having to take a week's unpaid leave, or, worse, it could lead to dismissal. East End families were mainly dockers, lightermen, leather manufacturers, costermongers, coal men or factory workers. Bermondsey women mainly worked in Hartley's jam factory, Peak & Freans biscuit factory near London Bridge, Atkinson's perfume producers, one of the various leather tanneries or at one of the street markets.

Say, for what were hop yards meant,
Or why was Burton built on Trent?
Oh, many a peer of England Brews
Livelier liquor than the Muse.
And Malt does more than Milton can
To justify God's ways to man.

A.E. Houseman

CHAPTER ONE

SETTING THE SCENE

Nowadays, people consider themselves poor if they cannot afford the latest television set or trainers, but the hoppers lived in an era when being poor was meant in its true sense; not being able to afford shoes for the children or food to put on the table. Even a day's wage was a significant loss to the family income, which is why any extra money, such as that they earned by hop picking, was vital as it was mainly used to buy essential clothing for the family, boots for the ever-growing children and a decent family meal for Christmas celebrations. Nowadays, it is usual to wear something clean and different each day and toss it in the washing machine once worn, but most folk had only their everyday clothes used for work and what was known as their 'Sunday best'. There were no washing machines until the late 1950s and only the well-to-do could afford them, so all washing was done the time-consuming way – by hand. Today, many elderly ex-hoppers suffer from corns and bunions because their parents could not afford the correct size of shoes for them when they were children. Footwear had to last until enough money was available to buy the next pair, and this could mean having to put up with tight shoes on growing feet for an extra six months. My mother, Doris Laverack Steeple, was almost obsessive about ensuring my feet were properly fitted with shoes as I grew, and I understood her concern as she suffered with painful bunions for most of her adult life due to having to wear tight shoes as a child.

Before the 1940s poor families lived 'on the bread line', meaning that they lived from one day to the next with their finances, never knowing if, after paying the rent, they would have enough cash left to buy bread, vegetables and a bit of meat for dinner for the family at the end of a hard working day. Tea was sold loose by the ounce in little triangular bags for those who could not afford a quarter packet. Eggs could be bought individually and dripping often replaced butter because it came from the joint of beef baked on Sundays. I think there is little tastier than a slice of bread and dripping with its rich brown jelly on a thick slice of homemade bread topped with a sprinkling of salt. Families were large, as many as twelve or more children, and it was a constant struggle to feed them all adequately.

Mash in a vat.

Hoppers having a well-deserved rest.

Home hoppers.

Hop pickers.

Within self-supporting families, the older ones looked after the youngsters. Many children did their bit to help top-up the family purse, paid for running errands for neighbours or doing other small jobs. In a valiant effort to boost his family income, Ron Amner not only ran errands at the age of eight, he searched the streets for cobs of coal dropped by a passing coal man or collected broken wood boxes from Woolwich Market before splitting them into small lengths which he sold as firewood at 6d a bundle. He also collected cabbage leaves, bruised fruit and discarded potatoes fallen off Woolwich Market stalls to have something extra to take home to his mother for dinner. He took pride in helping out financially, knowing that even a small amount of extra money was a great help to the family's comfort and subsistence.

The Woolwich house they rented did not have a bathroom and there was only one tap in the kitchen, for cold water only, which served the whole house. The rooms were generally chilly in winter as coal was expensive, and the family burned what wood they could find in the stove and fireplaces. If the nights were really cold the family piled overcoats and any spare clothing onto the beds, heated bricks in the oven and wrapped them in flannel or filled hot-water bottles to keep warm. Painful chilblains were common among the poor. Ron brought home any nuggets of coal he found out in the street after the coal man had passed by for his mum to use in the kitchen fire for cooking a hot dinner. Bedrooms, the kitchen and the parlour had fireplaces and, like most houses in the 1920s and 1930s, the only source of heating was by coal and wood fires lit in a fireplace which belched smoke into the room unless they could afford a chimney sweep. Their oven was in an old cooking range in the kitchen which needed cleaning out every morning, and so cooking and heating were both labour-intensive, time-consuming jobs. Ron's dad worked at the Arsenal in Woolwich and drove a train transporting cordite. Over the years he developed a sickly yellow pallor and eventually died from cordite poisoning, effectively leaving his grieving widow and children to get on with life as best they could. Parents and children wore the same clothes day after day because they had no other garments to wear, not because they were dirty people. It was a daily struggle to appear 'respectable'. Clothes were patched until little of the original material remained.

A SONG FROM THE EARLY 1900s

Little Mr Baggy Britches, I love you.
If you'll be my hopping feller
I'll patch them with purple, with green and with yeller,
And folks will say,
As we sit on the old farm wall,
'Someone been patching his britches
'Til he's got no britches at all.'

Courtesy of Doris Steeple

Pawn shops did a brisk trade even as late as the 1960s, and it was common practice for a family to pawn dad's 'wedding suit' (i.e. his best suit) on Monday morning to pay for the week's groceries, and then redeem it on Saturday morning after he had been given his wages on the Friday. So he had the use of his suit for the weekend but it earned them money during the week. Hopping was a dirty, clothes-tearing job and all the family's old clothes were carefully saved during the year and packed into the hopping box until August arrived, when they were sorted out. The ones in better condition were handed down to the next child, while the ones in poorest condition were kept for sharing and wearing down at the hop fields. Nothing was wasted, particularly gloves, which were needed against the scratchy bine stalks. Wise sayings abounded and were well heeded: 'Waste not, want not'; 'Neither a borrower nor a lender be', 'A penny saved is a penny gained'; 'Take care of the pennies and the pounds take care of themselves'. Such maxims were straightforward; saving was not. Every penny earned was needed for household expenses and any savings were usually kept in a communal pot, teapot or beer mug up on the mantelpiece above the kitchen range. It was seldom even half full.

It was a long walk to the village.

It was a matter of pride to never fall into debt. Family breadwinners were mainly men, but if the man of the family was in the armed forces during the war, or had died, it was the woman who had to earn, and women's pay was very low – usually about half the wage of a man doing the same work. That was one of the good things about work involving hop, fruit and vegetable picking – everyone got paid at the same rate. There was honour in work no matter how humble the job and, no matter how poor, most families presented themselves as well as their budget allowed. It was normal to have 'hand me downs', the clothes and boots passed down the family. In the hop fields we saw children wearing their father's old, torn and too large coats to work in the drifts, while the girls wore dresses and skirts with the hem lines tucked and sewn up to make them a better fit and an apron, often made from sacking, to protect their clothes. If they could afford any, it was normal for mothers to buy new clothing two sizes larger than required as the child would 'soon grow into it', which was true, but it meant sewing up hems and trouser ends so they didn't get worn out too soon, then letting them down an inch at a time as the child grew. Mothers knitted everything from vests to stockings and swimsuits (which tended to sag in the water). New garments in shop windows were just to be gazed at, not afforded. My mother was delighted to buy some off-the-coupon natural wool (with the lanolin still in it). She kindly knitted me several winter vests from it without realising how scratchy it would be – on a child's tender skin it was like wearing a hair shirt! My grandfather was in the Army in India at the beginning of the twentieth century and pay filtered down sporadically through the Army Pay Board to a soldier's family. In order to put food on the table for nine children, my grandmother, Emily, took in laundry, ironing, scrubbed floors and cleaned for those better off than herself and, despite long hours of hard work, still found difficulty in raising the 5s at the end of each week to pay the rent man for her back-to-back terrace house with an outside toilet.

Husbands of hopping wives only saw their family at the weekends if they were prepared to travel down to the gardens. In a time when few families could afford any kind of transport other than a bike, this usually meant a Friday-night train ride followed by a five- or six-mile walk to the hop farm from the station, often in the dark. The men were usually relied on to take down the weekend joint of beef when they went. Chicken was a luxury in the 1950s and was often only tasted at Christmas or 'high days and holidays'. Chickens were all free-range, as no one kept battery hens in the 1950s, so were more flavoursome than today's. Beef was the cheapest meat, and boiled beef, carrots, salted beef and silverside were popular meals. Herb dumplings, homemade gravy from the roast, doorsteps of bread and meaty suet puddings all helped to bulk out the otherwise frugal meals. It is strange that, despite all the fattening foods we ate, most of us didn't put on weight. I can remember regularly helping myself to a half pint of creamy milk from a dipper dipped into the farmer's milk churn and how delicious it tasted.

Hop picking involved wives being away from home for several weeks at a time, so their children had little choice but to go with them, consequently missing out on school. One of the reasons today's children have the privilege of long summer holidays is because, up until the mid-1960s, children regularly failed to attend the first two weeks of the September term due to extra picking, so when school boards set holiday dates they tended to allow for this long absence to some extent. If a family took on extra work after the hop picking had ended it caused problems academically for those involved, as the beginning of the autumn term is when new curriculum areas are introduced in schools. Hoppers recall that, as pupils, they were often forced to be absent at the time when the basics in important subjects such as maths or English grammar were initiated, and many found this impeded their work later in the year. Not all teachers were particularly understanding or sympathetic about this, and expected the pupil to catch up with extra homework or do as best they could. I remember my own maths teacher telling me, when I missed a point she had made in the newly introduced algebra, 'Well you should have been paying better attention, Hilary!', and moving on without further explanation or opportunity to revise.

One gentleman of eighty-nine years, Sidney Fagan, now with three Masters Degrees and a string of other qualifications to his name, told me that as a hop-picking child he was always behind in maths. Despite his later impressive qualifications, he is still haunted by those missed early weeks of the school term when he was in constant trouble for not being up to standard with his maths.

Education: The Way to Prosperity

I lived in Keatons Road, Bermondsey, as a child and my family, who were mainly lightermen on the Thames, together with the majority of our neighbours, were regular hoppers. Our house in Keatons Road and the local school took a direct hit during the Second World War and both were partly demolished. Fortunately no one was at home or in school at the time. It was some time before the school could be rebuilt. However, when the choice had to be made between extra money for a cash-strapped household, or improved education for the children it was the former that took precedence. Sometimes the older children were able to stay back in London with non-hopping relations or friendly neighbours which allowed them to still attend school, thus enabling them to catch up on the curriculum while the women of the family earned the extra money by hop picking. My parents saw a good education as the only sure way out of the poverty trap and did their best to get us children to school, but sometimes poorly educated parents could not see the point of 'wasting time' in school when a child's time could be more usefully employed earning money, therefore teachers trying to give children a good education did not always get the family support that was needed, unlike me. I particularly remember a colleague of mine being visited at school by the father of Vincent, who was struggling with his reading at ten years of age. 'I can't read and yet I run my own successful business, so if it's good enough for me, it's good enough for my son!' he shouted. Vincent's dad ran a scrap-metal yard.

Sidney Fagan

Escapism

For two to three months of each year the countryside was packed with the poor of London. According to the type of hoppers they met, villagers looked on the influx with either tolerance or dread. Some villagers willingly roasted a hopping family's Sunday dinner in their cottage oven for 1s a time, while others kept their saleable goods under key and wire netting when hoppers were in town, fully convinced that all Londoners were thieves, flea-ridden and mainly anti-social. At least, that's what we thought of them because the men always seemed to be fighting. Besides, in those days, anyone from outside the village was looked on as a foreigner.

Sarah George

Dr Salter to the Rescue

Socially-minded doctors, such as Dr Salter of Bermondsey, recognised the importance of the countryside's clean air, even though it may be for only a few weeks of the year. He ordered many of his patients with lung problems to be sent to work on the hop farms so they could breathe clean air away from the smogs of London. The word 'smog' is derived from a combination of smoke and fog, which leaves an impossible to remove soot-like deposit in the lungs; it was to be many years later before the Clean Air Act was put into effect, preventing people from burning smoke-emitting fuel. As a consequence of the Act there was a measurable drop in the number of people suffering from lung diseases, particularly TB (tuberculosis).

Sidney Fagan

Above: Dr Salter, a cartoon drawn by the author.

Below: A rough sketch map drawn by Ralph Salter.

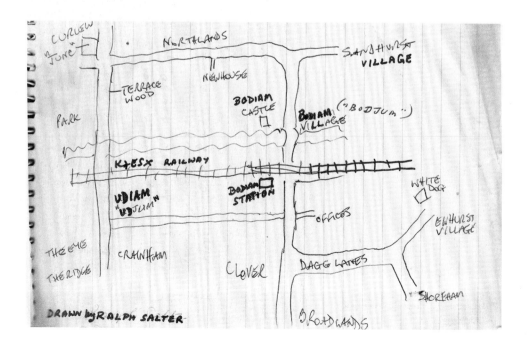

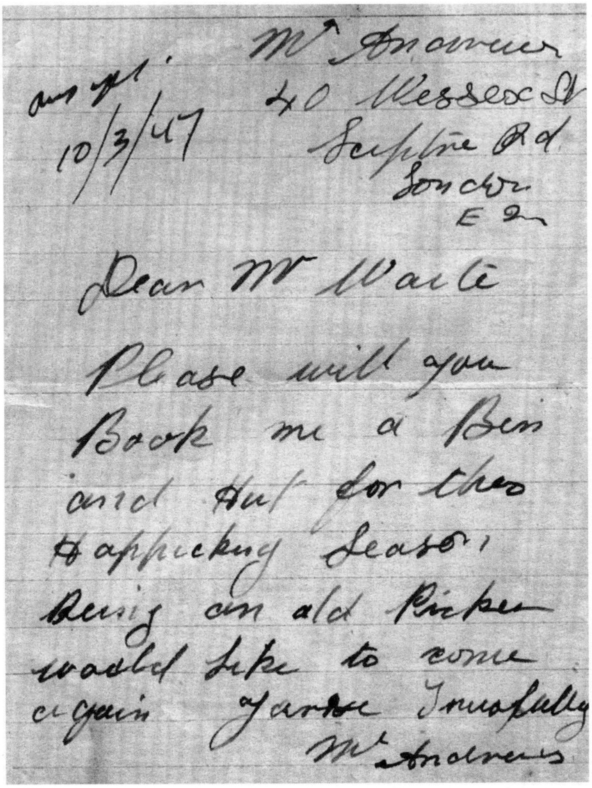

Mr [Answer?]
10/3/47
40 [Wessex?] St
[Scripture?] Rd
London
E 2

Dear Mr Waite

Please will you
Book me a Bin
and Hut for the
Happicking Season
Being an old Picker
would like to come
again Yours Truafully
Mr Andrews

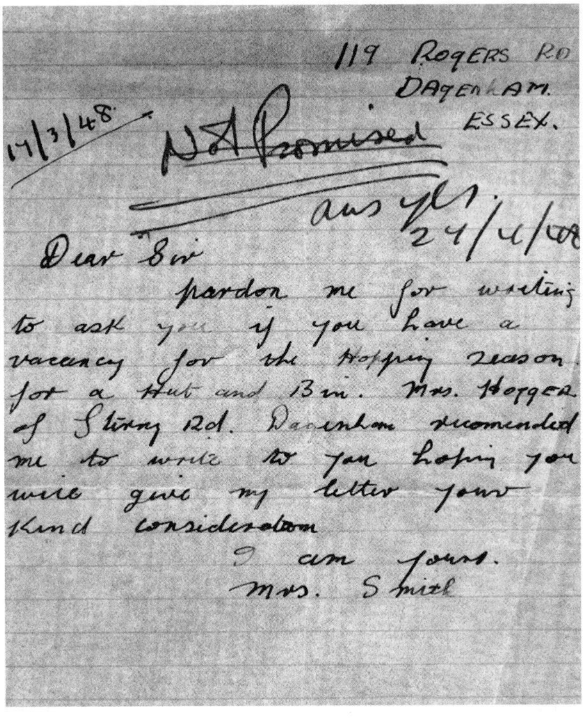

119 Rogers Rd
Dagenham.
Essex.

17/3/48   Not Promised

ans yfl
24/4/08

Dear Sir
        pardon me for writing
to ask you if you have a
vacancy for the Hopping season
for a Hut and Bin. Mrs. Hogger
of [Stainy?] Rd. Dagenham recomended
me to write to you hoping you
will give my letter your
kind consideration
            I am yours.
            Mrs. Smith

Above and below: Letters applying for hop-picking jobs.

Mrs. D. E. Bennett
75 Paradise St
Rotherhithe
S. E. 16.

5.2.48.

Dear Mr Waite
Could you
please book me for 1 Bin
and 1 hop hut for this
years Hop Season. I hope
you are all keeping well

ansyr.   Please Oblidge
7/2/48   Mrs. D. E. Bennett

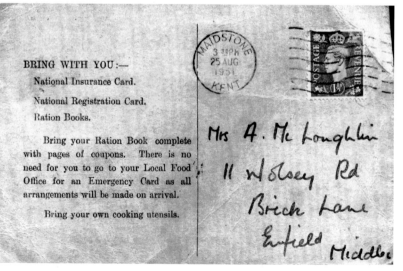

BRING WITH YOU:—

National Insurance Card.

National Registration Card.

Ration Books.

Bring your Ration Book complete
with pages of coupons. There is no
need for you to go to your Local Food
Office for an Emergency Card as all
arrangements will be made on arrival.

Bring your own cooking utensils.

MAIDSTONE
3.30PM
25 AUG
1951
KENT

Mrs A. McLoughlin
11 Wolsey Rd
Brick Lane
Enfield   Middx

We can go hopping!

CHAPTER TWO

TERRY AND JUNE HOGGER, FAMILY AND FRIENDS

Despite the hard work and basic conditions, families who failed to get a placement on a farm the following year deemed it a disaster in terms of the loss of extra cash and a means of escape from the harsh reality of city living. 'It was our only holiday. I was twenty-four before discovering that 'going on holiday' didn't mean working your socks off from dawn to dusk so you could earn enough money for new shoes and a slap-up Christmas dinner', recalled one ex-hopper wryly. 'It meant going to exotic places like Margate.' Yet for two months of each year the countryside was packed with the poor of London.

Terry and Rita Hogger were keen hoppers, as Terry's mother and grandmother had been before them. Terry's mother picked regularly over twenty-two years. Her children went down to the fields with her and she was fined 10s by the school board for keeping her youngest daughter off school. Such a sum was enough money for about fifteen loaves of bread in those days, and so Mrs Hogger didn't want that to happen again. 'She got around it by sending her younger children to the local school at Tudeley village while we were down hopping', said Terry, 'and the Board couldn't complain about that.' He remembers a far more serious incident:

Some hop pickers stole a car in London and drove it down to the hop fields to save having to find the train fares down from London Bridge. There was a fair bit of petrol still in the tank and they drove it all round the hop fields for about a week, lording it up like they owned it. But after a while the police caught up with them and they were both sent to jail for ten days to teach them a lesson. Hardly anyone owned a motor car in those days so it stuck out like a sore thumb whenever it went down the lanes to any of the villages so it wasn't hard for the police to catch up with them. My father could never have afforded to take his ten children on holiday so hop picking was ideal for us as it brought in a bit of cash and gave everyone a nice break in the country which we could never have done otherwise. There was always plenty to do and we children got up to all sorts of mischief, but never of the malicious kind like you get now. The owners of Postern Farm, where we picked, were Mr and Mrs (Mary) Waite and their daughter Susan, who still lives in the oast house next door. Sometimes we picked at Bugsey's Farm in Capel, but we mainly worked at Postern Park Farm which is just round the corner from the church in Tudeley High Street. It's still there and we go back to see it each year.

Left: Kathy Bailey and family. (The Hogger Family)

Below: Jacky Moody (née Hogger).

Clockwise from top left:
The Hogger men.

Mrs Betty Bird and young David Hogger. (Jacky Moody)

Stella Hogger and daughters Mavi and Marion in 1936.

Hop pickers. (Terry Hogger)

Hop pickers. (Terry Hogger)

Above: Hop pickers. (Terry Hogger)

Middle: Dinny Foley and her daughter, Lily, at Brenchley.

Below: The Foley family at Brenchley. (Lily Hogger)

Clockwise from top left:
Hop pickers enjoying a birthday party at Postern Park Farm, Brenchley.

Children in the hop fields at Postern Park Farm. (Lily Hogger)

Mrs Kate Hogger at Selling. (Jacky Moody)

Postern Park Farm, Tudeley.

Hoppers at Postern Park Farm, Tudeley.

Top: Postern Park Farm, Brenchley.

Middle, left to right: A friend, Terry Hogger's father, Terry, Terry's brother Jim, another friend.

Bottom: Terry Hogger.

The Foley family enjoying teatime at Postern Park Farm. (Lily Hogger)

Terry and Lily Hogger at Brenchley. (Lily Hogger)

Left: Fooling around at Brenchley. (Lily Hogger)

Middle left: The outdoor toilet.

Middle right: 'When we fell out of a tree'. (Lily Hogger)

Bottom: A family gathering at Postern Park Farm, Brnchley. (Lily Hogger)

Terry picked at Postern Park Hop Farm every year with his family when he was a child:

It was owned by Mrs Waite, who is now ninety-six years of age. She doesn't run it as a hop farm now but she still goes out into the fields and pulls the hop bines from the hedges around the local fields, carries them back to the farm and decorates all her rooms by hanging the bines across all the pictures. It smells just like being in the hop fields again.

Hopping in the 1930s at Postern Park Farm, Tudeley, Tonbridge, Kent

My mother, Kate Hogger, along with her mother, Granny Connell, and also her sisters, were regular pickers at the farm. We always had two bins and two hop huts because we were a large family. As soon as we received our hopping letter, we would book the lorry to take us down to the farm. We also picked up other hoppers who were also going down to our farm and then the cost could be shared by all the families. The first day would be spent in cleaning out the hut, making up the beds, filling the palliasses up with straw (that was what you slept on) and making the hut clean and tidy and not forgetting the hanging of the curtains. Once the picking was due to start, the farm bailiff would call all to work at 7 a.m., but before you started picking on the first morning the farmer, John Waite, would read out the rules and regulations which you had to adhere to or else you would be asked to leave the farm. Local tradesmen would call round such as the milkman, greengrocer and the baker who would sell us lovely doughnuts. The local pub, the Red Cow, had a garage which was turned into a shop for the hop-picking season and it sold everything from paraffin to postcards but you had to have your ration books with you as they couldn't sell you any food without them. We picked the hops into bins. The measurer would come round twice a day and take the hops out of the bin in a bushel basket. They were counted into a poke by the pole puller. The pokes held ten bushels. The picked bushels of hops were then recorded by the tally man into your record book. Then the hops were taken to the oast house for drying. My dad, Alfie Hogger, always had the first two weeks in September for his holidays because he loved hop picking. We always had visitors at weekends. Brothers and sisters, aunts and uncles – they all came to visit. The local pub, the Red Cow, always charged 2s deposit on a glass just to make sure you returned them. By the time September was over hop picking was finished and we all returned home. But we were already looking forward to next year's hop picking. Now it's 2008 and we still visit the farm that holds so many memories. We love to look out across the old hop fields and point out where the huts used to be. Happy days! Mrs Mary Waite, the farmer's wife, is still living at Postern Park Farm with her family.

Terry Hogger

The whole family helped with the picking.

We caught the train to Paddock Wood

I'm trying to think of things I can remember about the hopping, but there was so much. I can't find many of the tons of photos we had because they were lost when we moved but I have found a few for you. I remember when we all used to go to London Bridge to catch the train to Paddock Wood. When we got there we were picked up by the farmer's cart. Our mum always had the baby's pram with us which when we came home was always packed out with all the fruit we had got scrumping (stealing). Later, of course, we all went to Kent on the back of one of my dad's lorries which was exciting then, although the smell of the Woolwich ferry was rotten and we all had to cover our mouths and noses from it. I remember all the singing our parents done as they were coming back from the pub in the evenings over the weekend. And we (the kids) would all be sitting around the fire, baking spuds. And when we expected our dad to arrive on a Friday evening we would all go racing down through the orchard to meet him. I remember the funny taste on our fingers from the hops when we ate our sandwiches. The horrible toilets were just a hole in the ground with a wooden seat over the top with corrugated iron surrounding it. We had to walk to the tap across the orchard to get water and we had to go about a mile to get faggots for the fire. And oh the spiders and beetles and other insects – what about all those different types and colours of caterpillars on the hops. We were always the intruders as far as the locals were concerned. And one year we were chased by the man who guarded the fruit. He used to shoot the birds and he'd fire at us as well. Then there were the rabbits and the rabbit snares and the lovely stews we made from them when they were caught. Living so close to everyone else in the huts, as we did, you knew all their problems because the walls were only wooden slats and when Nan Clifton snored we shook two huts along. I enjoyed the fruit picking. We all had our little baskets on a belt round our waists and as we filled one we placed it on the ground as we all went along the line of fruit bushes. Sometimes some rotter would nick some but not the really regular

A hop hut at the Museum of Kent Life, Cobtree.

A farm cart at the Museum of Kent Life, Cobtree.

A hop-bagging press (*left*) and milk churns and butter pats (*bottom*) at the Museum of Kent Life, Cobtree.

Collecting bines using the new-fangled hop-picking machine (*below*).

pickers. We would often be ill because of eating so much fruit. We were having to sleep on mattresses filled with straw for many years until our dad brought our mattresses down from home. I remember the prisoners of war, men working on our farm and us being told not to have anything to do with them. Also I vaguely remember the fighting in the sky during the war when it got so bad some of us came home. I hated the getting up so early in the morning for hop picking and when the leaves were wet, how they smarted your arms. Sometimes we were put in the hop bins by the boys and this was awful along with all them beetles and spiders, but it was a happy, free time for us kids. We were scruffy and dirty most of the time because we enjoyed climbing all the trees and making camps in them. And the ropes swings from the branches, which many of the kids could never do at home in London as they only had small yards at the back of their homes. Cooking on open fires, boiling up your tea in a billycan, going into Tonbridge Wells for a hot bath at the bath house – all happy memories.

The lovely walks and views, and the different vans that brought different foods round to the fields, and the Church Army coming regular with films for us all to watch. As well, there was the fun we had when our parents got together and there was a football match or a tug-of-war team to compete with the locals at the pub, the 'Hop Bine'. At night we kids all slept together in one big bed. It was hard work at times, but as you look back it was a really free, happy and healthy time.

Lily Hogger

When the Picking Machines took over

The farm where we went hop picking now had machines to pick the hops instead. After twenty-three years of picking at the same farm we were upset to say the least. Our mum was given the address of another farm owned by Major Berry, who farmed at a place called Selling, near Canterbury. When the hopping letter arrived we were all highly delighted. The green hopping trunk was already packed with all our needs so we were surprised when, on the Friday night, a lorry called at the house to collect the luggage, followed by a coach the next morning. The coach went round picking up all the hop pickers to take them down to the farm. We never had heard of so much luxury. When we arrived we found that the huts had electric lights, and we were to pick in bushel baskets. Me and my brother still had to go to school. I went to the local school but he went to a school in Faversham. The reason for us going to school was our mum was fined 10s for each of us the year before by Dagenham Borough Council because we had been kept off school and missed our lessons. Saturday nights we all got the train from Selling Station to Canterbury, where we went to the White Hart pub. The governor of the pub would flick the bar lights on and off at ten o'clock. That was to tell us it was time to leave to catch the last train back to Selling. On arriving back we would have a huge bonfire and stay up to the early hours.

Jacky Moody

Rules and Regulations for Binmen and Pickers.

(Approved by the National Farmers' Union (Kent Branch) and the National Union of Agricultural Workers).

BINMEN.—Wages to be   per day, or in proportion thereto, for the time he may be employed. To be in the hop-garden in the morning at the time appointed, and not to leave the hop-garden during working hours without consent. To furnish hops to   bins. To look after the pickers and require them to pick their hops well and to see that all hops are picked up from the ground. To get all heads down which may be left on the wires or poles. To attend to and assist the measurers at all times. To assist in moving the bins. To carry off the pokes and load same on waggon whenever required. To assist other Binmen and not to use the pokes of another Binman without his consent. To take care of the bins, cloths, pokes, and other property committed to his charge, and to deliver up the same in good order at the end of the picking or on leaving; in default thereof to pay the cost of replacing what may be missing or unfairly damaged. For any breach of these regulations a Binman may be discharged immediately. Binman's wages to cease at the commencement of any strike.

PICKERS.—All pickers to pick the hops well and to their employer's satisfaction, and to be subject to the regulations herein set forth, and after the tally has been set they shall remain until the picking is all finished. To pick up all hops each time the bin is moved and to pick hops clean from strings and poles and free from bunches and leaves. To have their hops ready for the measurer when required. To be in the hop-garden and remain there at the appointed hours. For every breach of these regulations to forfeit one basket of hops. **The tally to be set during the first week of the picking and any picker who shall leave after the tally has been fixed for any cause whatever before the picking is finished, or who shall be discharged for conduct not in accordance with the foregoing regulations, or for other misconduct, shall be paid off at half the agreed rate subject to the Agricultural Wages Act.** In the event of any person wishing to leave through illness, a local Doctor's certificate must be furnished. Pickers may draw on account on such days only as shall be named at the commencement of the picking, up to an amount not exceeding one-half of their earnings.

GENERAL REGULATIONS.—Signal to be given by blowing a horn or otherwise when the picking is to commence or leave off. No hops to be picked during dinner time. No red-headed matches to be used within a distance of five hills from a bin. No smoking allowed near the farm buildings and stacks. All camp fires shall be extinguished by 10 p.m. No pokes to be taken out of the hop-garden by any picker or any unauthorised person. No pickers allowed in orchards or plantations. For any breach of these general regulations, offenders shall be liable to instant dismissal.

☞ Any person found stealing fruit or other farm produce, or damaging farm property, or committing any other illegal act, will be prosecuted.

FREMLIN'S

HOP-PICKERS'

Account Book

NAME ....

GARDEN ....

Published by
W. H. SMITH & SON, LTD.,
10/11 High Street, Maidstone.

Above: The Hop Pickers' Account Book for Bushel Court.

Left: Hop pickers preparing dinner.

CHAPTER THREE

LIVING THE LIFE

Life down in the country was relaxed and easy compared with manual work in the city, or so the visiting hop pickers thought. Nevertheless, especially on the larger hop farms which could employ as many as 4,000 workers, there were rules and with so many workers it was essential that these were followed in the cause of safety. Town and city dwellers did not always appreciate the real reasons behind this, not realising how dangerous some machines and areas of a farm could be, and rules were frequently flouted to dire effect because of this lack of knowledge.

Bin men were really the police of the hop gardens; they commanded a responsible position as they were in charge of ensuring that rules were not broken. They had to keep order among the pickers and if a strike or a serious fight broke out which he did not bring under control, his wages were immediately forfeited. One man could be in charge of ten aisles and as many bins needing hop bines for pickers to pluck. All the bins in their aisles were well supplied with bines ready to be picked. They had to ensure the bins were filled with hops and not leaves and twigs, ready for the measurer. When an aisle had been picked clean it was up to the bin men to help lift the heavy stake and canvas bins to the next available aisle, ready for the pickers to do their job. They were also responsible for seeing that the bins and other equipment, such as pokes, hop dogs and poles, were kept in good order. The replacement value was deducted from his wages should anything be missing or damaged. In default, he was up for dismissal.

Pickers also had rules to follow. Pulling hops off a bine and throwing them in the bin was no sinecure. They had to be picked 'to their employer's satisfaction'. This meant that once the measurer had been round and the tally man had taken and recorded the tally, pickers must remain at their bins until the measuring was finished in the garden and only then could they leave. Under the watchful eye of the bin men they were required to pick up all hops dropped on the ground around the bins so the area was kept clean. Should pickers leave or be discharged from the job, whatever the reason (other than by producing of a doctor's certificate) before the hop garden was completed, they would be paid for their personal tally at half the rate set by the Agricultural Wages Act. For

Above left: Hop-pickers' huts at Brenchley. (Lily Hogger)

Above right: A tally man.

Left: Inside a typical hop-picker's hut at the Museum of Kent Life, Cobtree.

every breach of regulations their tally was forfeited one basketful of hops. They were allowed to 'sub' up to half of the amount earned to that date, but only on a specific day of the week designated by the grower/owner. This was usually on a Friday or Saturday to enable families to have cash to buy groceries for the forthcoming week. Some families fared better than others in this regard, because if a parent (or both) delved into their joint earnings to pay for drinks at the pub to any extent, the money soon disappeared, even though beer was only 1s or 1s 6d a pint.

Picking had to begin once the foreman sounded a horn, whistle or other device, and this was blown again at dinner time when all work had to stop. The meal was eaten where they were picking and, except for calls of nature, no one was to leave the field. The whistle sounded a third time for recommencing work and the sweetest sound of the day was the foreman's call of 'Pull no more bines' because once the bines already stacked by the bins had been picked there was no more work that day and the hoppers could return to their huts for dinner. The most responsible child of a family was usually despatched back to the huts once the call had been made; their job was to get the fire going ready for cooking the evening meal if it had not already been cooked the night before.

There was not a ban on smoking while on the job. Back in the 1950s, and before that, no one understood the health dangers of the 'weed' and the majority of working men (and many women) rolled their own cigarettes as a cheap alternative to buying a packet. If they did buy commercial cigarettes it was often Will's Woodbines, sold in paper packets of five. Loose tobacco was carried in a tin or leather pouch in which were also kept a small packet of Rizzla or other papers to roll a pinch of tobacco into a thin cigarette. It was one of the few pleasures workers could afford. However, fires were an ever-present danger on any farm, and because hop bines are easily combustible the rules stipulated 'no red-headed matches to be used within a distance of five hills from a bin'. Hills are the mounds of earth surrounding a hop plant and were approximately 8ft apart, so smoking was not allowed closer than 40ft from the edge of the hop garden. The 'red-headed matches' were sulphur-based and capable of self-igniting, although they were safer than the phosphorous heads used prior to them. No smoking was allowed near the farm or oast houses, farm buildings or haystacks. All campfires were required to be extinguished by 10 p.m. and this usually meant damping down with water so no sparks could be blown into a flame should the wind get up in the night. Other rules were strictly applied. For instance, no pokes were allowed to be removed from the gardens by any unauthorised person – a poke of freshly dried hops was a valuable commodity. Pickers were not allowed in orchards or 'plantations', i.e. crop fields. 'For any breach of these general regulations, offenders shall be liable to instant dismissal. Any person found stealing fruit or other farm produce, or damaging farm property, or committing any other legal act, will be prosecuted', was the unequivocal statement in Fremlin's book of rules, but then half the fun of scrumping was that it was forbidden in the first place!

An abandoned tractor.

Twenty-Two Years Picking on the same Farm

I was born on 27 December 1914, just after Christmas and during the first months of World War One. I worked on the same hop farm for twenty-two years and when I started it was all poles, then string. The tally man gave out metal discs to the hoppers with numbers on them to denote the number of bushels they had picked. When the measurer called out 'Get your hops ready!' it meant he was coming round with his bushel basket to see how much you had picked and he recorded the amount in your hop book which every family had to keep the score. It might be eight or nine bushels he counted. Before he'd start you had to pull out all the leaves and bunches of hops you'd put in to bulk up the amount picked or, if you didn't, you got a telling off. At the end of the season you took this book to the office and they paid you according to how many bushels were marked down. But people had to take care not to lose it as, without proof of how much they had picked, they couldn't collect their money at the end of the picking. The hoppers slept on straw-filled pillows and straw-filled beds. After a few nights you got used to all the prickles from the sharp straw but by the end of the season there was hardly any bounce left in the mattress because the straw had broken up and gone flat. The hop huts were about 10ft wide and 6ft long, so not a lot of room for a whole family. We had to fetch water at the pump in a bucket for washing, drinking and everything we needed, so hoppers had to carry a lot of buckets from the pump to the hut, especially if they were for a big family. But it was lovely to smell the camp fires and the cooking but as I told you, I put up a lot of the hoppers in my house as it was better for them to be with me than in the hop huts. At least they could have a decent bath instead of just a wash down and lick and polish.

Alice Heskett

A lady with a lively mind, Alice was always out in the Kent countryside looking at new things and, once she retired, was a regular passenger on coach tours once around the country. One such tour was to Pontins holiday camp and the coach driver was intrigued by his passenger's extensive knowledge about the hops and countryside they were passing by and enjoying the scenery of Kent and Sussex. 'On one journey we were passing through hop country and the driver commented on this, briefly explaining what hops were.' Alice took it much further and started recounting her own experiences in the gardens. The driver was so interested in her stories of life in the hop gardens he asked her to tell her tales to we other passengers, so Alice settled down with the microphone and regaled us with numerous, fascinating tales. 'Instead of being in awe of the mike and her fellow passengers like a lot of people would be, she spoke with clarity and professional detail', said the coach driver, 'The whole hopping scene came alive for her listeners and everyone was charmed by her talk. She was a plucky lady and when we reached our next farm destination she boldly offered to sit on an ostrich! When I asked her if she could ride ... well of course she could! She was a real country girl. She was a truly lovely, genuine person', the coach driver told us.

Pole Taxing

My husband Ron and I both worked at our local farm when the farmer was preparing the gardens for the next hop crop. Ron's regular work was at the farm and when he wasn't working in the hop gardens he was working in the orchards. It was mainly apples but we had a few rows of pears and cherries as well so he was never short of work to do. We were given the job of putting up the wires and strings on top of the poles so we both had to go on the stilts. We had a tied cottage on the farm down by the road, that is, the farmer owned it but we could live there as long as we worked for him. As soon as we moved on, or if Ron had got fired, we'd have to be out of the cottage the next day. Because I was Ron's wife and we lived on the farm I was expected to work whenever there was need. I didn't get paid a wage as such because it was all down to Ron's wage as was the tied cottage, so in a way I was working for nothing. It was hard work and at the end of the day your legs really ached. I was always a bit scared of falling over because the stilts were about 12ft high and you really hit the ground with a right thwack if you lost your balance. Ron did the hardest part, pulling the wires along from one pole to another and I did the easier part, slipping metal hooks onto the wires to fasten the strings to. Ron did that second part too. The balls of string were about 18in across, about 35cm. They were heavy, too. He had a sacking apron

Above left: Margie Foley climbing a ladder. (Lily Hogger)

Above right: A stilt walker attaching new strings to the top wires of the hop string.

tied round his waist and the ball of string went into the pocket. He was on the ground for this part. He'd walk down the row with a pole with a hook on the end and threaded the string along the hooks we'd put up, then pull it fairly tight down to the heel and peg it into the ground with a piece of twisted wire which stopped it from moving off. Next we'd plant three bines at the foot of the string, as close as we could get them because otherwise they got caught by the tractor and pulled out as it went along and then you'd have a hop bine missing. Around March the bines were beginning to grow nicely and we'd go along nipping off all the pipey ones, that is, the weaker ones on the crown. Then it was me and the other women who went twiddling, or training the bines up the strings so they had something to cling onto as they got taller and heavier. Our farmer always insisted it was the women who did this job because he said we were less likely to snap off the growing tips than the men. After that we eased the bines away from the plant so they had room to grow, and then stripped off the leaves halfway up the stem because this let the air in and they needed that to grow and if you didn't it would grow really thick with leaves. I found the work very taxing because my knees and back ached from all the bending, and my hands got torn on the backs from brushing up against the plants and string. My fingers were sore from handling the bines on the rough string and I usually got quite a few nasty cuts on my palms and wrists. I couldn't wear gloves, really, although most did, but I didn't because I couldn't feel the tendrils with them on and sometimes I'd break them off just because I couldn't feel them properly. If you broke off one of the tips the bine didn't grow properly and you could be half a bine short by the time the picking came. We started about eight o'clock in the morning before the sun got up too high in the sky as it was much harder with the heat and no shelter from the sun. We knocked off about five, but it was a long day.

Amy Smith

Dirty Pickers

One thing our farmer couldn't abide was dirty pickers. I don't mean folk who needed a wash, because we all needed that after a day in the gardens. I mean people who threw the leaves and a few twigs into the hop bins just to bulk their pickings up a bit, hoping they'd get more money for less hops. It was cheating, really, because the leaves and twigs weren't any use to the farmer and his crop and the measurer made them pull out all the leaves and unnecessary bits before he'd credit them with their pickings in their hop book.

Margaret Barnes

Trials of a Stilt Walker

My Uncle George was a stilt walker at Whitbread's and his job was to tie the strings to the overhead wires strung across from the tops of the supports so the bines had something to grow up. The overhead wires were nearly 30ft high so the only way they could reach to tie off the strings was by walking along the rows on stilts. We had a real palaver when dad wanted to put them on as he had to climb up a ladder first, then step onto the little platforms on the stilts and fasten them round his shins and thighs with leather straps. He was about 10ft high or even more off the ground. If one of the stilts got stuck in the mud he'd let himself come down like a felled tree because it was too dangerous to grab the horizontal wires; they would have sliced his hand open. My dad, James Weekes, did the hop drying and was in charge of the ovens. The men who did this couldn't leave the fires once they were lit as they had to be kept at an even temperature while the hops were drying and dad was responsible for keeping the fires going and checking the temperature gauges. They used to sleep on slats with hop pokes laid on them like mattresses. They stank of hops when they came home for dinner. He lived at Ladingford and worked at Rookers Oast.

Shirley Hales

Make do and Mend

We were married when I was eighteen and we started off our married life in an old, condemned cottage that was infested with rats. We picked hops at Tong Farm, Benchley. My first mats were made out of hop pockets. I unpicked the side and bottom seams so they would lie flat, then dyed them a dark green after binding up the edges to make them look neat. They looked really nice on the sitting-room floor. That was over fifty years ago. When we were children we played on the hop pokes which were stored out the back of the oast house once they had been filled, waiting to be taken up to Borough Market, London Bridge to the Hop Exchange. One day I had a sore place on my back and when my brothers asked what was wrong I said I had a bruise. They told our dad and he saw I had a tick on my back and it was burying itself under my skin. Ticks are quite big and they bury their eggs in you or an animal if they get hold of you. My dad got it out with the tip of his penknife but the head stayed under my skin and it had to be disinfected out.

Daphne Chapman (née Baldock)

We hopped every year down at Keelands Farm near Paddock Wood. We were a big family and my mum was allocated four huts to accommodate us all. Mum found someone who owned their own car and trailer and paid them to take all our things down to the farm for us. Keelands had a proper purpose-built cookhouse so mum didn't have to get wet when it rained and she was cooking our dinner. Most of the pickers on our farm came from Peckham way.

Sylvia Barnes (née White)

Caught in the Rain

Our family were booked to go down hopping but when it came to Saturday morning I didn't want to go. The rest of the family got us to go down with them and once we were there, persuaded us to stay overnight and pick. I said 'Alright', and the next morning we

An engine on a 'Hoppers' Special' train.

picked for four hours. Unfortunately it was raining really heavily and we didn't have any spare clothing with us even though we were absolutely soaked by lunch time. When it was time to go home we were wet through from head to toe.

F. Ted

Scrumping in September

My Uncle, George Baker, used to pick us up in his removal van and carried fourteen of us, adults and kids, down to Whitbread's Hop Farm at Paddock Wood. Seven Mile Lane was a rough road then, and we'd get bounced around in the back of the van when he drove down it. As soon as we arrived my brothers would make a beeline for the orchards and nick as many apples as they could carry. Later that night my mum would take up all the blankets, carry them to the orchard and lay them on the ground all round the Damson plum tree, then she told us to sleep there while she went off to the village pub with the rest of the adults and we children wouldn't see her until she got back to the farm the next morning and made us breakfast.

Gladys Baldwin (née Giles)

Catching the Hoppers' Train

My sister Joy, our mother, Auntie Claire, and her four children and I had to walk across London pushing two hopping boxes with all our things in them to London Bridge

Station to catch the hopping train which took us down to the countryside. We had to be there by midnight as the train was ready to leave by then. There were people everywhere and you had to hang on tight to your mum or you could get lost among all the other hopping families. A few years later, Aunt Claire had the use of a horse and cart because her husband was a greengrocer. The horse had a mind of its own and when it had gone a certain distance he thought he was going to work and it was a really hard job to persuade him to carry us as far as the station because he kept wanting to turn round and go the way he knew. We enjoyed the train ride but once we got down there we had a lot of work to do. 'You can go and get the faggots', mum told me and I thought she meant meat faggots, a delicacy in the East End of London, so was disappointed when they turned out to be big bundles of twigs and sticks. These were used as the base for our bed and we put straw-filled mattresses on top of them to sleep on. The straw was prickly at night. We had to share the toilets and they smelled terrible so no one liked using them, especially in the dark. After work some of the men would light a big fire and we'd all sit around it and sing. It was bright round the fire, but dark at night all behind us out of the firelight and hard to see as there were no lights. When the trucks came to collect the hoppers in the morning to carry them to the fields we all had to wear Wellington boots because the ground was muddy and when we pulled down those first bines we'd get wet from the early morning dew and earwigs and bits of leaves fell in our hair.

Win Knighton (née Cordell)

Barrow of Beer

This man, I can't remember his name, used to sneak off the picking about midday and fetch his hopping box from outside his hut and push it down to the village to the pub. I think it was the Hope Pole, or some such. He'd order enough crates of beer to fill his hop box, pay for it, then push it all the way back up the hill to the farm. It says a lot for everyone in those days that he could leave it safely outside his hut without anyone trying to steal the beer, while he went back to the fields for the last hour or such for the day, then when night time came and we'd all had our dinner and a big fire was lit away from the huts, we'd all gather round and he's sell his beer with a ha'penny extra on a bottle to pay him for his trouble of fetching it the three miles from the village for us. At the end of the night we'd collect up all the bottles and stick them in his hop box ready for him to take back the next day. He'd get another ha'penny on each bottle returned to the landlord so he made a nice, tidy sum on a barrow of beer and we'd all been saved the three-miles walk after a long day's work, so everyone was happy.

J. Childes

★ Ha'penny – half a pre-decimal penny, worth approximately one eighth of today's penny.

Guiness in Clover

I can clearly remember the farming area where I hopped in Sussex and I'll draw you a map from memory so you can see where everything was – it's changed now, unfortunately. As a child I enjoyed exploring the area around the farm and Bodiam Castle and recall that the Crainham and clover fields were both for growing hops when my family were there. Now, they are used as open paddocks for ponies. A few hop gardens are still in the area but when I was still a child the whole district grew hops for the Guinness Brewery.

Ralph Salter

Feeding the Family

Food was always a problem with a big family and although we managed to get enough to eat when down hopping it was when we got back home at the end of our five weeks that there were difficulties because dad just didn't earn enough money to pay the rent and buy food as well, so as many of us children as could went out doing odd jobs and skipped school to run errands. But one year mum had a good idea and on the day we were due to leave for home she got my brother Harry and I to catch one of the hens running loose around the farmyard. When no one was looking (because they were all busy packing up all their own belongings ready to take home) we sneaked it into our luggage and hid it in the railway carriage. The hen wasn't very pleased about it and made a lot of noise. We only had two other people in the carriage with us but they were smiling so I guess they knew what we were up to. But, oh, didn't that hen taste good for Sunday dinner when we got home! It was dad who had to wring its neck, and it kept jumping around long after it was dead, which was a bit spooky, but when it was still, the girls got on and plucked it of its feathers while it was still warm, because they're easier to pluck that way. With a heap of mashed potatoes (we'd scrumped the potatoes the night before we left) and a good gravy made from the chicken bastings we had the best Sunday lunch I can ever remember.

George Green

Remembering Hop-Flavoured Boiled Eggs

My first recollection of the hop fields were back in the very early fifties. Waking up at 5.30 a.m. on a Saturday morning and waiting for Uncle Bill to turn up with his open back lorry (Uncle Bill was the rich member of the family). We loaded everything on to the back and then settled for the long journey from our flat in Bow to the hop farm in Sevenoaks, Kent. On the way down, passing and being passed by numerous others, all making their way to one of the many farms that was to be theirs for the weeks and, in some cases, months to come.

Arriving at the farm, the first chore was to clean out the corrugated-iron huts that were to be our living and sleeping quarters for while we were down there, then off to the main barn to pick up bales of hay that we used to fill our mattress covers for sleeping on. Next on the list was shopping. A mile walk to the local shop, lots to buy, which meant

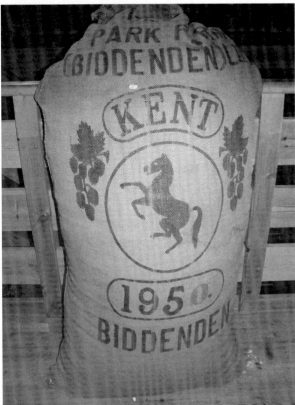

A pony trap (*above*) and a hop poke (*left*) at the Museum of Kent Life, Cobtree.

all us kids getting hold of paper carrier bags and marching off with nan for the weekly shop. Whilst we were away the men went to fill the gallon cans with water from the standpipe and collect the coal that was to be used in the braziers (empty oil drums with holes punched in them) that were used for cooking and heating when it got chilly in the evenings. With all the preparations completed it was time to meet the other families and rekindle friendships from previous years. Then it was play time because picking did not start until Monday morning. On Monday morning after early breakfast, off we went to the hop fields where we were allocated our plot. Row after row of hops strung up on their frames waiting for the pickers. The farm hands would come along and unhook the vines from the frames and picking would start. Our job was pulling the hop flowers from the vine and putting them into large canvas bags. As the bags filled up the checkers would come along with their lorries. Our bags would be weighed and our tally sheet signed. I cannot remember the figures but the more you collected, the more you earned. Under the blazing sun – and it always seemed to be sunny – it was not as easy to do the work as it seemed. A quota had to be maintained as there were always others who would take your plot. Under the watchful eye of grandad we all pitched in to make sure that our weights were up to the mark. Every so often grandad would call a halt, light his pipe and us kids would sit in groups, talking and laughing, exchanging stories of our schools, neighbourhood etc until the picking started again. Lunchtime would come with the arrival of Nan who had gone back to camp with some of the other women to prepare food. This was mainly salad with boiled eggs. To this day I can still recall the taste of boiled eggs with the residue of hops from my hands on them. Around 4 p.m. we would begin making our way back to camp to help get the evening meal ready for when the older members of the family returned. The smell of the food being cooked out in the open always gave you a whopping appetite and large potatoes cooked in their skins in silver paper on the brazier was a particular favourite. There was always someone with an accordion, and with grandad on the spoons the sing songs would go long into the night. Not too late, though, because there was always another day's picking to be done. Saturday night was the highlight of the week. We would all get ready and make our way to the local pub. A pint of lemonade and an arrowroot biscuit and us kids would be as happy as Larry. How times have changed.

The days just seemed to be filled with endless sunshine and laughter. Although it would be wrong to do it now, when we were children it was all right to go collecting birds eggs as long as you only took one from a nest, and many children had a treasured and carefully labelled collection from the hedgerows.

Tom Reddington

CHAPTER FOUR

A PRECIOUS HOLIDAY

Few could afford a holiday abroad in the 'good old days.' You were lucky if you could pay for a day trip to Brighton, never mind a month on the Costa. London pubs ran charabanc outings to the hop fields for their customers and folk paid 6d into a weekly fund until they had paid up the full amount being charged. The landlord running the outing usually chose a village where several of their regular customers were picking and it was always a gala occasion. The village pub they chose to visit was usually selected on the basis of it supporting the same brewery as the London pub patronised and the charabanc or coach often came home festooned in hops in honour of the day. The nearest to a comparable holiday today would be volunteering to help clear a river or canal of debris, or working as a volunteer at a National Trust or English Heritage property – except that for the majority of today's volunteers, there wouldn't be the desperate need to earn extra money, as was the hoppers' lot. Hoppers spent their holiday working hard in the open fields to earn their suntan, rather than sunbathing and opting out of work as we think of a vacation nowadays.

We Travelled in Style

We looked forward all year to going on holiday down at one of the hop farms. We took work wherever there was a place available and this could mean Yalding, Malling, Headcorn or Marsden according to which farmer was short of pickers. Some of the other pickers had to go down by train and that was always crowded, but we were lucky; we went down in my brother's lorry so we could just throw everything we needed in the back and that included our bedding, mattresses and the kids. There were a lot of us kids. Nineteen in fact. Sixteen of us went hopping but the others stayed at work near where we lived on the Isle of Dogs at 551, Manchester Road near the Queens pub. Mum went down in style with dad driving his car. 'Peace at last!' dad would say. We girls had to sit in the back of the lorry to go down. Once we got down to the hop farm we lived in three huts; one for the girls and mum and dad, one for the boys and the third one for living. The babies slept in their prams. We were usually up in the mornings 4 a.m. Everyone mucked in to help get us ready to go down to the hop fields. My older sisters used to dress the little ones and look after them, as was the way in big families then. There was a big brick open shed where we did the cooking. There was no time for us to play because there was so much to do. Because there were so many of us we were given four hop bins at a time. Father and the boys who were working all came down at the weekends and helped out. We worked from 5 a.m. to 6 p.m. so it was a long day. At night we all sat round the fire and relaxed for the evening, singing songs and telling stories. One of my brothers could play the flute and he played all the hillbilly music. My oldest brother was about twenty-four years older than me and the youngest, Maureen, was about ten months. It was the only holiday we ever knew. Even if the weather was bad and really raining we'd still keep on picking. The more you picked, the more you got paid. My older brother pulled down the bines for us and kept us well supplied so we had plenty to pick. He had to hook them down off the wires and carry them over to our bin, still on the strings. Generally we got six to eight weeks' work. The owners were Mr Knight, Mr Black and Mr White, whom we used to call Santa Claus because he had a large white beard. One day he chased me off with a stick because he caught me scrumping an apple. When it was time for dinner I loved mum's stews the best. Everything went in it but the kids. The farmer kept bantams and was pretty good to us as he gave us eggs and freshly baked bread and eggs most mornings, so we were well looked after and the fresh food was nice. We were as naughty as most kids and sometimes we'd open the big five-barred gate and let the horse out. This meant the boys got off doing anything because they had to go off and find it again while it was still light. Once we filled a bucket with leaves and sticks and tied it to a lamb's tail. We didn't mean any harm; we were just being mischievous. We were always late back to school because the hopping took so long to do, so we nearly always missed out on the introduction of a new subject we had to learn, and kept getting behind with our lessons.

Carol Goodman (née Wickers)

A Precious Bag of Sugar

It was wartime and everyone was on rations. This included meat, cheese, tea, sugar, jam, marmalade and butter. Even margarine was restricted and we used to mix the margarine and butter together to make it go further. Only bread, milk and vegetables were off the ration. You couldn't get many kinds of food without using coupons out of your ration book. When we were down at the hop farm Mrs Titmus ran the village stores, and although we knew she had lots of sugar (because she kept it in a pile on the counter where we could all see it), she would never sell any to the hoppers when we went to buy our groceries, so one day I nicked a 1lb bag off the counter as we went past. But that was nothing to my two brothers: they nicked the farmer's motorbike when he wasn't looking and took it off down the country roads. They weren't very expert and they had an accident, leaving the bike in a ditch. One of my brothers broke his leg doing it, so once the broken leg had been plastered up at the hospital, the farmer (who had to go off and get his own motor bike back from out of the ditch, so wasn't very pleased) made him sit beside our bin, gave him a whole lot of bines to pick and for the rest of his time there he had to pick hops for nothing as punishment.

Helen Wright

Hopping by Lucky Chance

We went hopping when I was a child and I particularly remember one man when I was in the hop fields one day and it was raining really heavily. He was wearing a bowler hat and it had a split in the top. The rain got in through the vent and water was cascading down through it and across his face but he still kept his hat on! We were evacuated by the government to Southborough during the war one August and we didn't know anything about hops or about picking at that time, but the farmer asked mum if we would all pick as they were really short-handed, what with all the men being away fighting and no one left to work the farms. So we were lucky and stayed there from September to December.

Roy Beaven

A Lesson Learned

When my parents first took me picking hops I was only about eight- or nine-months old. The hop bins they picked into were shaped a bit like a cradle and they used to stick me in at one end while all the adults hopped. I stripped off some of the hops and put them in my mouth to eat and they made me sick for three or four minutes. When the farmer came round to see how everyone was getting on the next day, they told him what had happened and he couldn't stop laughing.

Michael Cuffe

We Made our Own Fun

My mum got to bed about 9 p.m. as she was so tired, but we kids went to bed much earlier because we had to be up early to get down to the fields. There was my mum, Alfie who was ten, Georgie who was twelve, me, who was eight, and Lisa who was nearly two. Emma was only a few months old. Once at the hop field the littlest, Emma, was put in her pram while we were picking and Lisa, who was still a toddler, went in the bin for safety, as they couldn't get up to much mischief there and the sides were too high for them to climb out. The hop farm was quite remote and there was nowhere to go, so we stayed at the farm and we older lads had to make our own fun, climbing trees, playing football in the field with sheep, chasing each other playing tag, exploring the woods and hedgerows and making up 'pretend' games. There were lots of other children there so we always had someone to play 'chase' with, or make up a football team. In the evening we played ludo, snakes and ladders, rummy, fish, I spy, snap, beggar my neighbour, happy families or draughts. If one of the other families had a game we'd borrow it or lend them ours; everyone shared everything in them days.

Anon

The hop is such a pretty flower,
In shades of palest green.
The gardens with their singing sets
Make for a happy scene.
Yet in the spirit of this tiny bloom
Lies a secret worth the telling:
In its heart are tiny seeds
From which the oil is welling.
So innocent, it looks I know,
So delicate and dainty,
Yet the reputation of this flower
Is not, at most, called saintly!
When treated to a nice hot bath
And malted in the tun,
This little flower changes self
And decides to have some fun.
Unto every pint of beer that's made
A special essence it imparts,
And as the beer goes down each throat
There's gladness for our hearts.
'Drinks in, wits out'
You often hear folk tell,
So let a happy haze rule tipsy minds
Until we hear the bell.

Hilary Heffernan

CHAPTER FIVE

THE HOP'S STORY

Hop growing has been going on in England since the Romans brought hop crowns over from Italy in the first century and later travellers carried them home from the Continent, but the first book to be written exclusively about hops was *Trade of the Flemminge* printed in 1573. An English dictionary printed in 1140 noted one item: 'Hoppe. Seed for beyre.'

Only female hops are used in the process of brewing beer. There are many varieties, and names such as Fuggles, Phoenix and Boadicea come easily to brewers and growers alike. Several varieties such as Goldings are particular to Kent and named after a Kentish family. Some species are grown specially for their part in the brewing process; others are used to impart a particular flavour and a distinctive, earthy taste to the finished product.

Hops are strong, decorative climbing plants, which may grow as high as 25 ft, so need to be well supported. Many people, particularly ex-hoppers, grow them for decorative purposes in their gardens, and if left to themselves the bines will take over the nearest bushes, garden sheds or fences with impunity. However, it is not parasitic so will not kill off its host plant. The petals of its distinctive small flowers are similar in shape to a pine cone, and are properly called 'cones' but, instead of being wood, are delicate, almost translucent, and a pale green in colour. They look like posh little crinolines, said Pam Price.

The actual plant mass is called a crown and each crown is capable of producing over a dozen hop bines. Most of these are removed and only two, three or four bines are allowed to develop according to the method of planting used by the hop farmer. The premier shoots are delicate and can easily be snapped off so great care is needed when the plant is first sprouting. Women are usually given the task of twiddling, or persuading, the new shoots to follow the strings up to a stout crossbeam where a worker on stilts has fastened these to an overhead wire earlier in the year. Twiddling is an unpopular, back-breaking job and it can take a dozen women a week of working from 8 a.m. until 5 or 6 p.m. to complete a field, depending on its size. It is a labour-intensive task. The

Digging in the new hop crowns at Northiam.

usual quota was to twiddle between 800-900 shoots per day, and women new to the job were hard put to keep up with those who were more experienced and hardened to the work, which is rough on the fingers, arms and back and takes concentration while learning.

The old varieties of hopes grew naturally to a height of 25ft or more and were easier to harvest by hand, but new varieties developed since the 1970s usually do not exceed 12ft or so, and are more suitable for harvesting by automatic hop-picking machines.

The hop industry has its own vocabulary, as does any specialised commerce, and, unlike my old fashioned typewriter, I am amused to see my computer does not accept hop-related words such as 'bine', 'oast', 'scuppet', 'roundel' etc., (all terms well known to hoppers) without first putting up a fight – a bine is similar to a vine, only applies to hop plants.

The roundel is a conical, round building containing a hop-roasting fire which fast dries the hops to a specific temperature before they are scuppeted out onto the connected drying floor which is the long shed-like room of an oast house. A scuppet is similar to a shovel, with a pole handle, wood frame and scoop made of sacking. It has the advantage of not damaging the delicate hops when they are scooped up. While the hops are drying they need to be turned regularly to ensure they are thoroughly dried as it only takes a handful of damp cones to send a whole, tightly packed pocket (sack) of hops mouldy, so great care needs to be taken in the drying room to turn every part of each batch of hops. Alice Heskett remembered:

> Once the hops had been roasted and laid out on the drying floor the smell was so strong I couldn't stay there for long at first, but you soon got used to it.
>
> Hops make you sleep and there's many folk have made hop pillows as they couldn't get a good night's sleep without them. We even put the babies in the hop bins with the newly picked hops because the smell soon sent them to sleep and we were sure of at least an hour without them crying or us having to change any nappies. Now they sell hop pillows in some places and they charge a lot of money for them because they're in fancy pillowcases but we'd cut up an old sheet and sew it up, or even a bit of sacking and that did the job just as honestly.

Once the hops are dried they can be packed into the 6-ft long pockets by means of a bagging press which crams in the small cones solidly until full. This is then tightly sewn across the top opening with special white, waxed, coping twine (using a specially curved coping needle) and sent to be labelled in the stencilling room. The stencils display important information such as the name and village of the hop farm, the date the pocket was bagged and maybe the farm's own insignia and the destination of the hop pocket. Mr Doherty of Larkin's Brewery, which has been established well over 300 years, has a wide collection of such stencils showing how he and his family exported their hops abroad, even as far as Australia. During the Second World War brewing was considered an essential industry as millions of gallons of beer were exported abroad to our troops to help keep up morale.

Newly strung garden.

Three-month-old hops.

Hop cones.

Picking shed at
Little Halstead.

Newly
planted
Boadicea
hop crowns.

A scuppet.

A Foreign Language

If you were new to hop picking and had just arrived at your first hop farm, you almost needed a dictionary to understand what people were talking about. It was like a foreign language, what with scuppets and hop dogs and Bordeaux mixture. I remember hearing the grown ups talk about such things as 'wilt' which is a disease hops get such as split leaf blotch, sclerotinia wilt, and phoma wilt. There was Tutsham and Rodmersham, which are two kinds of the Golding varieties first recognised at Tutsham Hall in West Farleigh and the Rodmersham which came from Rodmersham Farm. Then there was 'cold bagged', which meant the hops in the pocket had been badly dried and could turn mouldy and a man had the special job of taking samples from a sack selected at random to see what the quality was like. The sampler was like a very large fork with either two, three or four cutting blades which were plunged into the side of a sack and a square sample taken out. If it smelled mouldy or damp in any way, that pocket was thrown out and another one or more had to be sampled to make sure we didn't send poor quality hops to the Exchange and damage the farm's reputation. Once you start thinking, all the old terms easily come back to you. Some of the stuff we handled would be labelled downright dangerous nowadays. I mean the Bordeaux mixture we used as a fungicide was made up of copper sulphate and calcium oxide – you'd know that as lime. The copper sulphate is highly poisonous and the lime burns enough to take the flesh off a corpse. They used it to bury people in times of plague back in the old days. We didn't use gloves or masks back then either. Then there was Flowers of Sulphur. Not only did we use this as a fungicide on the growing bines, but sulphur dioxide was infused into the hops when they were being dried; we put rolls of blocked sulphur into a metal pan or shovel into the oven and as the heat got to it we let the fumes feed through the hops lying out on their horsehair mat up in the upper part of the roundel to give them flavour. There was a sort of safety area all round the edge of a hop field and this was there for several reasons. The ones I know were that it was a fire break in case there was a fire and the break gave us time to stop it reaching the hops, which burnt easily if a fire got to them. It was also to allow the various carts such as the nidget and the shim (both kinds of hoeing implement) to get quickly to a part of the hop field, and it was where the hoppers often went to eat their lunches. Pedlars and local trades people came down it to sell their wares to the hoppers, such as the bakers, butchers and the fish man. It was also the limit of when the hops were sprayed against blight or wilt as well as fertiliser. Knowledgeable hop pickers tried to avoid being allotted the drifts at the edges of a hop field because the hops there were often smaller; the fertiliser didn't always reach the outside rows, you see, so didn't do so well, and it needed more small hops to make up a bushel than big ones.

George Green

Working the drifts – a watercolour by the author.

Kent oasts.

Converted oasts.

CHAPTER SIX

WHERE HAVE ALL THE HOPPERS GONE?

Most ex-hoppers will tell you that the real death of hopping came about when mechanised pickers were brought in to do their job.

The decline of the hop gardens began in the 1960s when it became more economic to import pelletised hops from Belgium and Germany, but the real death of hopping came when mechanised pickers were brought in to do the job. Where a garden may have employed 1,000 hoppers to pick their fields, they now required a mere five or six people to work with the mechanical picker and, particularly over the past fifteen years, hopping has faded to little more than a memory. Sarah Green said:

> If they started it again I'd be there tomorrow … it was really bad news when the farmer got us all together at the end of the season in 1963 and told us we weren't going back again next year because he didn't need us after that season because he had this wonderful new machine and he wouldn't need us pickers any more.

Whereas some farms formerly needed to employ anything up to 4,000 or more hand pickers for a season, one of the new machines could deal with a whole hop garden employing a dozen or so workers when using the machines. This meant the farmer's overhead costs were vastly reduced. He no longer had to pay an army of workers by the bushel for the hops they picked, supply hopping huts, faggots for their beds and fires, straw or hay for their many palliasses and put up with having his orchards stripped of good fruit on which he relied to make a profit at the end of the season. With a hop picking machine he only needed to pay for the initial outlay for the machine, employ half a dozen workers to control it, a couple more for each oast's furnace, a couple of baggers and a few men to move the pokes and pockets around as they were filled. At a stroke, the farmer cut his costs mightily and thousands of country-loving hop pickers lost their prized part-time jobs and their holidays.

The first machines were massive affairs which needed a building the size of a barn to house them. As designs progressed, they are now reduced to something about the size of a

An early automatic hop picker.

Hop-stripping machine shed at Little Halden.

A picking machine at Northiam. (Iain Ross-Macleod)

Above left: The Column at the Hop Exchange, Bankside, London.

Above right: The gated doorway at the Hop Exchange, Bankside, London.

tractor and need only one driver to operate. In accordance with the new style of picker a new style of hop was developed and the so-called 'dwarf' varieties evolved. One of these is the Boadicea which grows to a mere 12ft tall, is easy to pick by machine when in full bloom and is a speciality hop favoured by the Rother Valley Brewery which, like many specialist breweries, grows its own hops. At one time hop farms were allied to one brewer or another, for instance those in north-west Kent sent most of their crop to Shepheard Neame, who claim to be Britain's oldest brewery. Mid-Kent farmers sent their hops to Whitbreads, who owned one of the largest hop farms at Paddock Wood. Sussex farmers sent their hops to Guinness, who owned farms around Bodiam and Eastbourne.

Progress is an invidious thing and once the industry discovered that hops could be pelletised, or freeze-dried in much the same way as instant coffee, the demise of hop gardens started in earnest as it became more economical to import the pellets from Belgium and Germany.

Faversham, Maidstone, Lamberhurst, Bodiam and Five Oaks Green were some of the major hop-growing areas, but there were many smaller hop gardens in south-east England, reaching as far south as Sussex and north to Canterbury. Worcester and Herefordshire were two other prolific hop-growing areas and it seems that the soil that is good for fruit trees is also excellent for hops, as all these areas are also first-class orchard country. These farmers usually sent their hops up to the Hop Exchange in Borough High Street, London Bridge. The impressive size of the Exchange is indicative of just how important hops were considered up until the 1980s.

The Hop Exchange at
Bankside, London.

The Hop Exchange portal.

Looking into the Hop Exchange, Bankside, London.

Hop growing was far more widespread than is realised today: For many years I taught at Alderwood School, Eltham, which I later discovered was built on the site of a former hop garden. Eltham, now included in Greater London, was originally in Kent and is a mere twelve miles from the capital. (After his half-brother, William I, invaded Britain, Odo was given Eltham as his due and built a manor where Eltham Palace now stands.)

Now 'hopping' has faded to little more than a memory. 'If they decided to start up the hand picking again I'd be there tomorrow as the first volunteer in line', say many enthusiastic, but displaced hoppers. Nell Hearson was a hopper who lived in Welling with her plumber husband from when they were first married in the 1930s. Nell had six children and they all went down hopping with her:

In those days plumbers weren't paid a fortune for their jobs like they are nowadays. My husband got a set wage each week and it wasn't really enough to feed a big family, I can tell you, and mine wasn't half as big as some of them. I never 'subbed' my hopping money like some of them, so by the end of our time down on the farm I was able to collect all I'd earned from the clerk who paid us off at the end of the season, and it meant we had enough for a great Christmas with presents for all the kids and a Christmas dinner I was proud of. Dad used to come down to see us at the weekends and he always brought down a piece of beef or a few trotters or chaps of pork, which were the cheaper cuts in those days, and I'd put them in a pot over the fire and by about 6 p.m. it would be done to a turn with a few carrots and potatoes scrumped the night before to throw into the pot with the meat. I tell you, those meals cooked in the open air were the best I ever knew. You could make brawn from the chaps, and that made tasty sandwiches for our lunches. Most pickers are now in their sixties to nineties and still recall their hopping days as some of the happiest of their lives.

Irene Crimmins, her sister and son Richard and the rest of their family went hopping for many years at Goudhurst: 'We loved it, going down to the farm every year,' said Irene, 'The smell of the hops, all our friends and neighbours down there with us, that lovely smell of cooking in the open air, this was what we looked forward to every year.' In 2006, now in a wheelchair and unable to go hopping any more as their old farm had moved on, Irene visited me at the Hop Farm Country Park (formerly owned by Whitbreads Brewery) for their hop festival every year. Sadly, Irene died in August 2003. Her ashes were scattered on the hop field where she had hop picked for so many years and her son Richard had a very special floral tribute made in memory of his mum. It said 'Gone Down Hopping.' God Bless you, Irene.

Kindly Remembered

I was a friend of Irene Crimmins for over sixty years. We used to pick together at Carters Farm in Ewehurst Green. The war was on while we were there and when the German aircraft came over our dads used to throw us under the hedges so we couldn't be seen. But we could watch the dog fights going on overhead between the Spitfires and the Messerschmitts. One aircraft was shot down and we saw the pilot bail out and come down in our fields by parachute. It was always the same families that went down

Marion Healey's father, Thomas King, when he was aged seventeen in 1924, wearing a cap.

to the farm, year after year, so we all grew up together and it was lovely meeting people again: the Jones and the Heads. We had a birthday the same day as Tommy Head and we'd have a birthday party in the middle of the common round a big camp fire and we all sang songs. It was lovely. Our bed was made of faggots covered with a palliasse stuffed with straw. It wasn't very comfortable but we got used to it. The last time I hopped was in 1952. I must have pushed my two little girls miles in their pram! My grandparents lived up in Bodiam and dad and mum used to walk there to meet them at the pub for a pint.

Rose Thompson

Local Hoppers

We lived all our lives in Southfleet and when the hopping was on, my mother, Mabel King (my dad was Thomas King), took us along to Northfleet Green, only a few miles away, and we would pick for the day then go home in the evening. We picked from 9 a.m. to 5 p.m. I think in other places they mainly had Londoners down to do the hop picking but our farm only had local people in to do the work. We children were sat in the hop bins and given sprays of hop to pick and once they were finished we threw the stalks out and left the hops in the bin with us. My older sister, Stella, now Stella Child, was at work so she didn't come down with us as she had to go to work at her job and is currently a local historian with the Bexhill Museum. Mavis Pavitt, who is two years older than me, picked with mum and me but when the twins, Barbara (Hodges) and Brenda (Green) were born, mum stopped going to the hop fields as she had so much to do, especially when later our brother Stuart was born because he had Down's Syndrome and died at twenty-eight, so took a lot of looking after. But we enjoyed hopping and we were always

Under the bagger.

sleepy sitting in the hop bins. I remember we took our lunches along with us but it was nice to have a good dinner at home.

Marion Healey

Vizards Farm

We lived in Tonbridge and in the 1950s my mum took us all hop picking. It was always a good day out in the hop fields. I was only small so mum sat me in the hop bin so I couldn't get up to mischief. I can remember watching the farm tractor pulling along a trailer with all the hop pokes piled high, off to the oast house. All the hop fields have gone now, which is sad. It's funny but the weather always seemed to be fine when it was hop-picking time. We picked at Vizards Farm, Haseden, near Tonbridge.

Gillian Harris

CHAPTER SEVEN

The Hop's Story in Wartime

Hard, Seasonal Work

Most workers today would expect to be changing jobs pretty soon if they were being strafed by aircraft machine-gun fire as they worked or in danger of a few tons of downed Messerschmitts landing next to you in the field where you were working, but pickers during the thirties and forties took such 'mishaps' in their stride, and continuing picking after dealing with the problem. People tended to be more resourceful in a time when so little help or goods were available and not so inclined to depend on others. As part of their war effort they were prepared to face an armed enemy pilot with little more than a stick or pitchfork to defend themselves. Work went on, regardless.

Doodlebug Danger

When I was a kid during the war I was down hopping with my family when the dog fights were on overhead. You could see the Spitfires in the sky above us and they were coming over to shoot down the doodlebugs which you could hear as they flew over our fields. The doodlebugs had jet engines and gave out a strange whistling sound as they flew. They said that when the whistling stopped the doodlebug was about to fall and explode. One day I had gone to the chemical toilet which was in the middle of a field and didn't have a window. As I sat there I could hear the Spitfires shooting as they flew overhead and then there was a really big bang. I was caught, really with my trousers down as I thought the toilet was about to be hit and I ran outside with pooh all down my back and legs. I was right – a cannon shell fired by the Spitfire had hit the bucket I'd been sat over and it had exploded in the bucket. I was glad it was only the bucket that was hit and not me.

Terry Abbot

A Natural Trap

We had finished picking one field of its hops and all that was left were the poles and overhead wires. We'd just moved on to the next field when there was a loud roaring noise

Digging a sheltering ditch against air attacks.

and an aircraft came down just above us. It was a Messerschmitt ME109 and it landed in among the bare wires. The pilot slid back his windscreen and climbed out onto the wing looking very shaky. The hoppers surrounded his plane holding any bits of wood or implements they could find so he couldn't escape, but suddenly 'Dad's Army' arrived and surrounded the aircraft and the pilot. At first the pilot was going for his pistol but then when he realised how many hoppers and Home Guard there were he thought better of it, and he and the Home Guard captain saluted each other. The Home Guard marched him off and the aircraft was left up on the wires, which made a natural trap for any enemy aircraft trying to make a landing.

Anon

Tackling the Enemy

We picked at Whitbread's farm when it was owned by the brewery. We were still picking even when the war was on and one day when everyone was out in the hop fields, there was a dog fight overhead and our lads shot a Messcherschmitt down. It landed in the hop garden and some of us went over to see if the pilot was still alive. He was and the Home

Guard took him prisoner and led him off the hop fields. We came down to Whitbread's to work for six months of the year every season, as we didn't just pick the hops, we also did all the early year work such as planting out the bines, twiddling them up the strings and nipping off the spindly shoots or 'pipey bines' as we called them.

Mrs Thomas

Nearly 100 and the Memories are still Bright

I was in the Air Force for twenty-seven years. I joined up with them in 1926. I'm ninety-six now (in September 2006) and came down to pick at Whitbread's every year as a child. I'm one of your oldest hoppers, now.

'Tommy' Atkinson

Agricultural Leave

My dad was in the Army but he tried to get his leave at hop-picking time and he'd come down to our farm where we were working where he'd work at pulling the bines, shifting the pokes and took his turn with the other men at the heavy work. He'd stay as long as his leave lasted. He'd be wearing his uniform and I must admit I showed off a bit in front of the other lads when he turned up looking all smart. Then at the end of his leave the farmer told him, 'You've pulled your weight, right enough and if you like, I'll get you Agricultural Leave for about six weeks and you'll be able to stay here.' Whether or not he got it depended on whether his regiment was about to be shipped abroad or not as the war had priority, but the farmer gave him a letter saying he was needed at the farm and his senior officers gave him permission to carry on, so he had an extra six weeks with our family. We finished the hops, went onto tatties next and then did a bit of fruit picking.

George Green

Streets Overflowing with Hoppers

You should have seen them, they used to walk in their hundreds from all over London, to get to London Bridge Station with all their hop boxes and stuff. I worked right near the station and could see them all arriving from my office window. You'd never believe there was enough work for all of them.

Barbara Moore

Those Bonfire Nights

We all picked at Sandhurst hop fields every year and I've been going since the age of eight. The part I loved was the big bonfire the men lit at the end of hopping before we all had to go home. We children were all given toffee apples to eat. There was a brick cook house with a brick chimney. The cooking fires were a couple of 5-gallon oil drums with the ends cut out, placed over a dug-out pit. They'd put a couple of bars across the opening to support the huge cooking pots. Most people cooked stews as it was the easiest and quickest – we were all hungry after a day's picking. Some of the fathers cycled all the way from London to be with their families at the weekends.

The Smithers family

Love in a Tent

My mother came down hopping at the Guinness Hop Farm at Sandhurst, Sussex. A whole lot of us travelled down together; that was my mum, me, my two kids and neighbours. We all came from Curton House, Canning Town. We went up Shooters Hill on the back of the lorry and it was a lot steeper than it is now so we all slid to the back of the lorry. One young couple who were newly weds chose to live in their own tent rather than one of the hop huts. As they didn't have electricity they used a hurricane, or storm, lantern inside the tent and didn't realise that when it was dark the lantern threw their silhouettes onto the tent wall and we could see them getting undressed and being all lovey dovey.

When it came to the weekend we bought a large joint of beef for the family dinner and I put it into our cooking pot and stuck it into the camp-fire embers to cook overnight and weighted it down with a couple of bricks. In the middle of the night somebody's dogs came round the camp looking for food, found my joint of beef and stole it so we'd no meat with our dinner the next night.

There is supposed to be a ghost at Sandhurst. They say that a lady had been drowned in the pond there and some nights her spirit would come out of the water and float around in the air because she was still devoutly restless as she'd fallen off the humpback bridge and drowned. The humpback bridge is still there, but I don't know about the lady spirit. The children used to catch rabbits in the fields and they made us a tasty dinner when I cooked them. They tasted a lot better than today's rabbits and poultry but they lived naturally then, not in wire cages There were lots of them around in the early morning. I had two children and I was given half a bin to fill when we were picking. It was hard work when the kids were little but then I took a young girl down with us to help by looking after the children and that made things a lot easier. We accidentally set the hop-hut curtains alight with the Primus stove one night and there was a bit of panic while we were putting the fire out. My husband came down at the weekends on a 'pop pop bike' and it had a hard seat and made his bum sore. What we all liked best was being allowed to ride on the back of the tractor when it was driven up to the oasts. When the weekends came we would all walk down to the Hammer, and once my sister-in-law and her husband, my brother, got a car to drive down to Hastings to see the sea and the car broke down on the way back. We were really worried because we had to be back on the Monday morning to start picking again. Then this 'knight of the road' came along on his motorbike and ferried us back to the huts sitting behind him on the bike. When it was time to be paid for our work they opened a little window up at the farm and we had to show how many bushels of hops we had picked before we were paid any wages. My dad came out of the Second World War after being shot and couldn't get a job until 1949 so we were glad of the hopping money. We put up net curtains at our window and door at the hop hut and a homemade mat on the floor to make it cosy. My son took his first steps down on the hop farm, so he was a very young hopper. The children picked into an apple box because the sides of the bins were too high for them to reach. The freedom was lovely and children had the freedom to run around safely, rather than in the streets of the city. The hop poles made natural goal posts and, when they had done their bit of picking, the children played football in the empty part of the hop fields which had already been picked.

Edith 'Edie' Gear (née) Saville

Terence, this is stupid stuff:
You eat your victuals fast enough;
There can't be much amiss,' tis clear,
To see the rate you drink you beer.

A.E. Houseman (1850-1936), *Last Poems*

Harvey's Brewery

CHAPTER EIGHT

SHARING THE MEMORIES

These stories are hopping history as remembered by the people who were in the hop gardens at the time. The events they record and the processes described are as they recall them so are not necessarily historically accurate.

'We are the sum of our yesterdays.' Reminiscences and recollections are important. They are the mementoes of our past and prove to ourselves that we have lived a life worth having. They tell us who we are. I would always encourage people to write down their life's story. After all, it is unique to them. We may well have shared large parts of our life with others, yet each of us will tell the story differently because within that shared memory are incidents that were unique to us. So if you have someone in your family who recalls lots of events from their past, encourage them to write it down, record their tales on a voice recorder or take notes and type it up later. After all this is part of your family's history and therefore part of you. Unless you do this, once the person has gone, so will their stories and there is no longer the culture of families sitting together, recalling the past so that their stories pass on down the family. Computers, so useful in their way, are put to other uses such as games, internet etc., whereas they are perfect devices for recording the stories and incidents told by our relations, adding to our family's incredible historical tapestry. I took it for granted that my grandparents would always be there to ask about their past and now many questions I would like answered are left to drift in the wind. Think about it. Don't leave it too late

Swig's Hole

I was about seven years old when we started going to a small farm at Horsmonden called Swig's Hole. The farmer allowed those who were booked in for working to go down a weekend before the season began to clean and paint their huts ready to move in when the hop picking started. I hated school so could never wait for the summer holidays to begin. My friends at the school and at the flats we lived in by Lambeth

Walk were envious when they stood watching my father's lorry being loaded with our belongings on the Friday night ready for the early morning journey down to Kent. His boss allowed dad to use the firm's lorry to take us down to the farm, then dad and his cousin, Mike, would stay until late Sunday afternoon and make the journey back up to London so they had the lorry back, ready to start work on Monday morning. There were always some empty huts where people went home after a few days because they could not take to the way of living, which was fairly rough living, a bit like camping. When our families or friends wanted to come down at the weekends the farmer was good in letting them make use of the empty huts to stay in. During the week, our hut was used by my mother, two brothers and myself. Dad and my sister stayed up working in London during the week but came down every weekend. Even close neighbours came down and it was like a little holiday for them. I loved those times. Everyone got dressed up on a Saturday night, even the children, then it was down to the village green where the grown-ups enjoyed their tipple and we children drank lemonade. That was the only time we were allowed to stay up late. Alas, like you wrote in one of your books, it all came to an end when the machines began to take over. I now live in Wales and have done so for the past thirty years. Now my family are scattered all around Kent. Each year I go to visit my sister and we go down to Swig's Hole Farm. Of course where the huts used to be is just grazing land now, but the old cookhouse that was used if it rained, and the old water standpipe were the only things we could relate to of the old times. The farmhouse was exactly the way it was when we were there and we spoke to the farmer who now owns it. As we felt awkward looking around, we explained why we had come. He was very nice and understanding. He told us that the farmer who had owned it when we were there is still going strong at the age of ninety-three, also his son owns another farm in the same area. He allowed us to look around at our leisure. My sister and I must have looked an odd couple standing in the middle of an empty field reliving our times there. It was really wonderful and something to look back on. When my sister bought your previous book she took a quick glance at it because she wanted to pick out a few faces she might know from our picking days, then she sent it on to me. I am writing a book about our family for all our children and grandchildren to look back on as these times will never come again and I think it is important they know how we used to live.

Pauline Ansell, Port Talbot

Old Songs and Old Flames

When it was a warm day and the sun was shining and steaming the early morning hops it was a lovely feeling. The smell of the hops was very strong then. We'd work fast and the hops soon piled up in the bins, everyone singing as we picked. Mum had back trouble so we used to take an old stool down for her to sit on while we were at the bins. Those evenings round the fire at night were magic when we thrust the largest potatoes we could find into the bottom of the blazing fire and they came out only partly cooked and black and they had to be split open but when you sliced them with a sharp knife and put a knob of butter and a bit of salt inside it was like angel's food. After a few beers we were all laughing and singing. Bart and his banjo were the best and Auntie Aggie sang the old songs, like I remember from the 1920s such as 'He was a real dirty dog, but so 'andsome.' I can't remember much of it but another bit of it was 'He told them that Santa Claus died in the night and so all the poor children got none, not a sausage …' When we sat by the fire, if there was a wind blowing, your backs got really cold while your shins were blotchy red from the flames. It got so hot it scorched your face and bare legs and there was a friendly crackly sound as the logs burned. Sometimes there was a noise like a gun going off and that was the flames finding some resin or gum in a log and exploding. You could see green and blue flames in among the yellow and it was from the resin. Someone would pass round a few pockets (hop sacks) and we'd put them round our shoulders to help keep warm. The night times were usually chilly if the weather was bad and the hop huts were draughty so I'd fill my old stone hot-water bottle and tucked myself round it when I went to bed.

Nell Hearson

Hop Dogs

A hop dog was shaped like a crocodile's jaws and was used for levering poles out of the ground. We still have one! The long tool with a curved blade and a spike is a hop hook; this was used for cutting the bines at ground level. The curved side was used for pushing the bines up and off the pole. Both these tools were used in our hop garden. I am sending you a copy of a hop-pickers account book. The rules are interesting.

Nora Batt

Pammie Price went hopping with her family as a child and loved every minute of it. 'You don't know your alive until you've been down the hop fields', she used to tease me. Pam ran a double sweet stall and a family shop in Deptford Market, London, until she died in 1999 in her late seventies. Many of the market traders were ex-hoppers and although hop picking was over at their old gardens they hired a coach every year to go down to the Gun at Horsemondon. Sometimes they got as far as their old farms but mostly stayed on the green outside the pub and talked about old times, tucked into their bangers, liver and mash dinner and enjoyed a pint or two. One year Pammie arranged for me to go with them. 'As I promised', she wrote in her letter, 'two tickets to meet real people, very good for your writing.' Pam recalled the time when hop tokens, given out to represent the number of bushels they had picked, could be used as currency in some village shops or pubs. 'They were as good as handing over money' she explained, 'because the shop keepers knew the farmers would honour them.' Her family continue to run the sweet stall in Deptford Market and I'm sure Pam is still around not far from hop heaven.

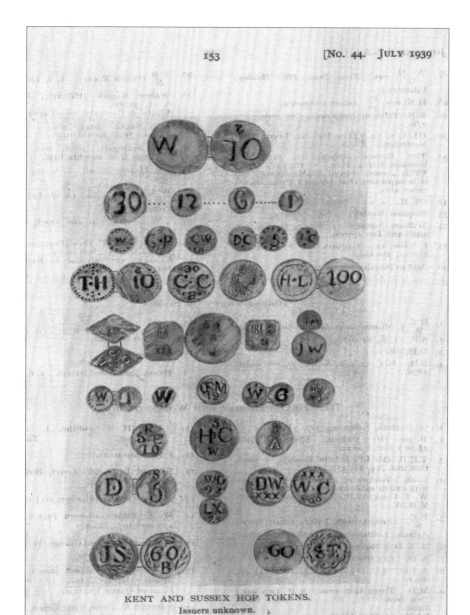

KENT AND SUSSEX HOP TOKENS.
Issuers unknown.

Hop pickers' tokens.

Old Signs for a Drinking House

Hops weren't the only fruit used for beer making. Not many people know about it but once they made beer from the red berries of the wild service tree. The leaf was a bit similar to the hawthorn. The old name for it was the chequer tree and it was called that in Roman times. It became 5 per cent proof after fermenting for only four days so it was a quick, easy beer to make at home by the country folk. No sugar was needed. You just covered the berries with water and yeast and in five or six days it was ready to drink. The name chequers goes back to before the Roman period. The Egyptian God Osiris used the chequer board (draughts) as his sign and the Egyptians and Romans used it outside their drinking houses as a sign that ale or beer could be bought there. A service tree was often grown outside pubs and so you got the Chequers Inn or the Chequers Tree. Most people couldn't read or write in those days so they used pictures people would recognise so they would use a horseshoe outside a blacksmiths or a boot outside the cobblers shop. A follow-up to that is in the nineteenth-century *Dictionary of Fame and Fable* by E. Cobham which states:

Chequers. A public house sign [and is part of] the arms of Fitzwarren, [who, in the days of the Kings, Henry] … had the power of licensing vintners and publicans [and was a sign that] the house was duly licensed.

But the sign has also been found on houses exhumed in Pompeii, the Roman city that was buried in the ashes of Mount Vesuvius across the bay from Naples. There is a Chequers Inn in Deal, Kent, and I'm sure readers can tell me of many more public houses named so. Around Kent and Sussex are many pubs called the Hop Pole, the Hop Inn, the Hop Yard and similar names, all referring to the hop industry and many can lay claim to being their hoppers' local.

Lennie Speers

Joys of Scrumping

We often scrumped in the farmer's orchards. My aunt made toffee apples for us with them. They were easy enough to make once you could get the apples. She made toffee over the camp fire, melting brown sugar in a pan, then caramelising it before sticking a peeled twig into the top of an apple and dipping it into the hot toffee. Then she'd stand the toffee apple on its end in a baking tray and as it dried, a lot of the toffee ran round the apple's sides so that when they were ready each of them had a nice lump of crackly toffee at the top of the apple. After we had eaten our share she would fill the tray up with toffee apples she'd ready to take home after we'd finished the picking. Her boyfriend had been down picking with us and they fell out on the way home. He was so angry he threw the tray of toffee apples off the back of the lorry onto the road so we didn't have any to give out when we arrived home. We picked at Reeves Farm, Sponden Lane, Sandhurst.

Anon

Binding the Hops

At one time they didn't use twine to fasten bines to the poles you know. No, they used dried rushes of flag leaves. Flags are tall, wild irises which grow in marshy land and have what I call 'happy yellow flowers'. We had to dry the long leaves gently after splitting them into strands and this was a hard job because the edges of the leaves were sharp and I was always cutting my hands. The rushes were much easier to work with. Before we set to work tying up the young bines we'd put the reeds or flags into a bowl of water and leave them to soak overnight. That way they would be easier to bend around the bines and strings and tie off. The next morning we women would grab as many as we thought we'd need for the day – usually from six to ten handfuls according to how fast a worker you were – and stuffed them in the pocket of our sacking aprons ready for when they were wanted, because we didn't want to waste time having to walk forwards and back to the farm to get more when we could have got enough in the first place. The water from the reeds used to soak through my skirt and then my legs got chafed as the damp material kept brushing against my thighs. Sometimes my legs were red raw by the end of the day. Some of the women carried the reeds in enamel bowls or jugs or put them in sacks, but it was easier not to have to drag something around with you but just to have them fastened at my waist. We had a little Valor fire in the hut and it got hot enough to boil the kettle on the top of an evening, or sometimes we'd light it because it was really cold some of the nights. We had to fetch the paraffin from the village and it was heavy to carry in its round, metal can which had a screw cap. One of the children burnt herself on it one time and after that we fixed a metal box around the stove with wire so it was safer. The Primus made it nice and cosy in the hut, though. It used paraffin and had a wick that had to be trimmed or it smelled really sharp if the wick was blackened.

Nell Hearson

The Quiet Comic

Barry was my hubby for over thirty years before he was taken bad and died of infected lung problems. He was a tall, sad-faced man but he had a secret side to him as he was a man with a sense of humour and in the evenings when we'd had our day in the gardens, picking all those smelly hops, we'd all gather round a blazing fire the men made up with wood from the copse behind the huts, and after dinner they'd all tell stories to us as we roasted scrumped tatties in the fire. Barry would come out with these serious-sounding comments and it was a real scream. You couldn't help but laugh, looking at his long face as well. I can't remember many of what he said now, but one of my favourites was 'You women are always complaining you never know where we husbands are at night', but he'd say with a cheeky grin 'I know at least one wife who always knows where hers is for sure – it's Mrs Smith – and that's because she's a widow.' Or he'd say things like 'When the boss said I should go far, my mate Bert said "Yes", and the sooner he clears off, the better!' or 'Ernie took his whole family hopping every year and was always complaining about his wife's mother. He

A partly made barrel.

said she was so ugly he took an instant dislike to her because as he knew he'd never be able to get along with her so it saved the time.' With that long face of his he had everyone in tucks. He was very popular with the other men and they'd all have a laugh together while they were pulling down the bines with their hooks. That was one tough job and his arms always ached at the end of the day because he'd got an injury in the war and it affected his muscle. The other men used to sing in Chorus once one of them started them off and that was usually Bert. He was a tenor. Barney had the deepest voice and he'd sing our favourites such as *Old Man River* that Paul Robeson sang. The singing really made the work go quickly.

Doris Ryman

American Hops

We grow hops in Oregon and Washington State over in the States, but also you'll find them being grown in Seattle, San Francisco and Nevada. We tend to have microbreweries over home.

Bill Carter

George and Elsie Maddock

We were married for over fifty years. I was apprenticed to be a carpenter as a lad and made shives for barrels and wood spiral staircases for a living. 'A shive is the proper name for a barrel bung,' my boss explained. Several farmers made their own barrels, and there was one particular hop farmer who had his own coopers to make barrels for his beer and I made the bungs to stopper his barrels. Being wood, I tapered them on my foot lathe. I had to stand up to do it but you got used to it. The shives were easy to fashion and fit because once they were set into the barrel and it was filled with water (to test if it could hold the beer or ale without leaking) the newly shaved wood quickly soaked up the moisture and expanded to fit the barrel hole, making it leak proof. The majority of shives I made were about ¾in thick then I put them on a rotor machine which made a ring cut in them. I could do several hundred in a day and they'd be delivered to the hop farmer by horse and cart the next day.

George and Elsie Maddock

Both Elsie and George visited the author about once a month for many years, then Elsie died in 2006 and George followed her soon after in 2007.

Pearlies, and a Well-Earned Pint

Two other regulars who picked at Eridge, near Tudeley, were Margaret and Brian Hemsley. After a day's hard work in the gardens they took Margaret's mother and old Bert down to the Tudeley local pub for a pint of an evening. The Hemsley's are now Pearly King and Pearly Queen for Harrow.

Many 'Pearlies' were regular hop pickers and one was Nancy Armitage, the Pearly Queen for Whitechapel. She went back to their old hop farm a few years ago after arranging a nostalgia hop-picking weekend for the family, staying at the George pub in the village. She was disappointed to find that only one of the many hop huts on their farm was left in reasonable condition. The rest had been allowed to fall into decay.

I still remember hop picking at Caring Farm near Leeds in Maidstone, Kent, when I was very young. My grandfather, Thornton Skinner, was the grower. I couldn't stand the smell but my wife says she loved it. Her family had hops, too.

John Snow

The Problems of Keeping Clean

We'd get in from the hop gardens all dirty and scratched and needing a bath but there wasn't one and the best we could do was to boil up a kettle of water, pour some of it into a white enamel washing up bowl and cooled it down with cold water, then we'd strip as far as was decent and given ourselves a good wash in the basin. It did the job all right and got the dirt off but there wasn't much we could do to get the brown hop juice off our hands and by the end of our time there we had really brown hands. The boys in the hut next to ours used to pee on their hands and they reckoned it got rid of the staining. It wore off after a couple of weeks once we were home, anyway, so we didn't really bother. Mum didn't like it and she tried to clean her hands using cucumber juice but it didn't seem to do much good. She'd rub the cucumber on her hands and face, then of an evening she'd sit with her hands up behind her head because she said this made the blood drain out of them and she'd have white hands like a lady. We kids all thought it was a waste of time because we'd be back to hopping the next morning and her hands were as bad as ever.

Doris Ryman

Tea on the Hop

We had Red Cross men come up the fields in the mornings, calling out 'Mug-O' and we'd go and get a cup of tea from them for a penny. They sold tea and sweets. At the end of the day most of the children used to go to the big hut on the edge of the cornfield and have a cup of Oxo and play a board game for a penny. This was run by the Red Cross as well. I was the youngest of five girls. Mum and dad had two weeks' holiday with us at the hop fields then they would leave us with our aunts and uncles to look after us. But then they would come down at the weekends to make sure we were all right and do a bit of picking. We used to pick in the aunts and uncles' bin for them. Sometimes we would sub (that is we'd ask for some of the money we'd earned from the office). One year we had no money to come at the end of our time picking, because we had subbed it all. Another time was when the farmer's chickens went missing. He came round the hop huts looking for them, but my dad had already hidden them in my pram and I was sitting on them. We had them for our dinner that night. When we came home from hopping all the family had fleas in our heads and we had to go to a special flea clinic to have our hair washed in flea soap before we were allowed to go back to school. Our family lived at the Elephant and Castle and used to go hopping down to Reaves Farm about three miles from Hadlow. We had to walk into the village to do our shopping as there wasn't a bus. We had to carry all our water from the pump which was about a half a mile away from where our hop hut was. We took down all the spare buckets and jugs to fill but it seemed we always managed to get more in our shoes than we saved in the buckets before we got it all back to the hut. Sometimes we used a big stick to carry the buckets and jugs. We slotted the stick through all the handles then two of us lifted it up with one of us on each end of the stick. It was really heavy and we lost a lot of the water before we got back.

Mrs P. Frost

Sing-along Journeys

My daughter is seventy-five years of age now and when she was a child I used to take her hop picking. I'm ninety-eight years old now but remember all the things that used to happen. My name was Mary Laycock and I was picking before I was married. At the end of the hop-picking weeks it was all singing and all dancing. The Lolly Man came round with sweets for the children and he sold delicious toffee apples which were one of my favourites. We usually finished picking on a Saturday about the middle of the day and load all our belongings into the back of the horse and cart, which was how we got down there sometimes, pile in with all the boxes, pots and pans, then we'd sing *Down at the old Bull and Bush* and other musical-hall songs everyone knew, all the way home. Other times we went by lorry and one time it broke down and we were all left standing by the side of the road with our bedding, chairs, cupboard, wash bowl and all the other things we needed until they sent back to London and they brought another lorry down to get us to the hop farm.

Mary Abell (née Laycock)

A Large Family of Pickers

We used to come down to the Guinness hop fields in Sussex sometimes We had a big family. There were twelve of us and we picked at Paddock Wood and Five Oaks Green until the farmer sold out and it was bought by Redsills, so we ended up by going to pick at Faversham, Halstead and Redmans.

Jean Richards (née Clarke)

A Birthday Pudding

When it was my sister's birthday it was always hop-picking time and we used to make her a spotted dick for her birthday.

Violet Goodchild (née Pearson)

'Did you ever taste beer?' I had a sip of it once,' said the small servant.
'Here's a state of things!' cried Mr Swiveller ...
'She never tasted it – it can't be tasted in a sip!'

Charles Dickens (1812-1970), *The Old Curiosity Shop*

A hop wedding at the Hop Farm, Country Park, Paddock Wood.

CHAPTER NINE

THE WAITES FAMILY AT BRANDBURY
AND BARBARA ROBERTS' FAMILY

Mr Waites' family were keen pickers and looked forward to being driven down to Brandbury each year. He recalls some of the incidents that happened when they were down there very clearly:

We picked on Church Farm in the 1920s to 1950s. I was about four in 1924 when I first started. We used to hire a lorry between two or three families and we'd load all the things we needed: the tea chest which dad had made into our hopping box and other items such as a couple of chairs which the adults used to sit on in the back of the lorry and in 1924 the lorry we were in was an old Ford of 300cwt. And we'd drive down through Wrotham and over the bridge, then turn right onto the Seven Mile Lane, sometimes by the Five Mile Lane. That day we were all singing and hadn't a care in the world, but suddenly we were nearly upside down in one of the deep, roadside ditches and the lorry tipped over. If we had gone over to the left it would have fallen into the river. Luckily there was no one killed but some of us were injured and shocked, including the driver. We were going to Malden on Brandbury Farm, Mainwaring. I was born in Stepney and practically everyone who came from the East End – Bow, Bethnal Green, Poplar and Bermondsey – went hop picking. I recall when we got down to the farm, I remember the farmer used to scramble apples every Sunday morning for the children and he used to tip them out so we had to run behind the truck to pick them up. As regards the huts, they were wooden and some were galvanised. The wooden ones were limewashed and filled with straw so when we arrived we had to fill our beds (palliasses) with straw ready for the night. Also, the tea chests were used as our tables. When we went to work in the hop gardens hawkers used to come round with their goods and he had things for the children, too. One of the hawkers we called the Lolly Man and he used to sell all kinds of sweets to the children. That's how the parents got their children to spend time picking hops, by offering to buy them a lollipop if they picked a bushel into an old man's upturned umbrella, which held about a bushel. One of the first things we had to do when we first arrived at the farm was to get three stakes of wood, like fencing stakes, and bang them in the ground so we could hang our pots on them for the cooking. Two went into the ground and the third was fixed across the top so there was somewhere to put the 'S' hook to hang the pot, which was very big and black with soot from the fire. If it rained we had what they called a cookhouse with a galvanised roof. It had three sides with walls and one side open. There was a steel rod going right across the wall where we could hang our pots over the fire to cook our meals. I remember one of my elder brothers used to go down hopping on a pony and trap with my Uncle Harry because he was nervous after the lorry tipped over. It used to take them two days because they had to stop for the pony to rest and be fed. I remember one year when my Uncle Harry came down with the pony and trap and the pony got colic. Unfortunately the pony died and was buried in a field in Brandinbury Farm. All the farm hands got together to dig the grave for it. So, as we got older and our parents passed away, God rest their souls, we took their places on the farms they used to work on and then I went in the Army in 1940 and came out, demobbed, in 1946. In 1949 I started going hopping again. There are many more memories but I'll come to a close for now.

Mr J. Waites

My Grandmother, an Intrepid Soul, would try Anything

I only went once, but that made such an impression on me that, fifty years later, I can recall memories and minute details of that once-in-a-lifetime holiday, although my poor mum saw it as anything but a holiday.

My mum, dad, young sister and I lived in a Brighton terraced house. I was eight years old and my sister was three. My maternal grandma and Great Aunt Lizzie were regular visitors and spent many hours gossiping over numerous cups of tea in our back room. It was Aunt Lizzie who told us about hop picking. Grandma was an intrepid soul who had led a colourful life and was already ready to try anything once. Mum was the complete opposite, but was persuaded to give the 'hop-picking lark' a go, probably with the thought of earning extra money. Grandma arranged the whole thing for us. It was August 1950 when our family joined countless other pickers at the Boorman's Hoathly Farm, Lamberhurst, Kent. We arrived at the farm on the back of a lorry hemmed in on all sides by a conglomeration of cases, boxes, pots, pans and an assortment of humanity, all strangers other than my family. In a field were two rows of black corrugated huts between which were two to three bonfire sites. Many huts were already occupied by families from London who regularly hopped each year and whose huts held all the comforts of home. Furniture, cooking facilities and curtains were all left in place year after year as these families always occupied the same huts. We were allocated one of the spare ones, completely empty except for a bed framework with string and roped criss-crossed to make a mattress base. Mattress? We hadn't got one. My poor mum's face! The thought of this tin shed being 'home' for six weeks must have appalled her. She had been forewarned to sew two double sheets together, and she filled this pocket with hay to make us a mattress or pally ass – what grandma called a friendly donkey. Auntie Lizzie, grandma and mum soon had the palliasse made and topped with mum's pristine sheets – it was quite comfortable.

We had brought a small, methylated-spirits Valor stove, pots, crockery and pans and, with curtains loaned for the small window, it was soon more homely. I suspect it was more fascinating for me than for poor mum! The only method of cooking was on the stove and this also did for heating and for boiling water for washing, cooking and tea. How mum managed this primitive contraption I'll never know, but I don't remember going hungry. The only toilets were in a shed at the far end of the field, which housed a plank with a hole over a bucket set beneath it. The smell was appalling and I pity whoever emptied them. There were many children, who seemed quite a rough lot as they ran wild, climbing trees, paddling in streams and getting dirty. But there were no reprimands as they suffered from even more dirt and grime in London and the countryside was better than that for them. I never felt to be one of them and rushed crying back to the hut one day when some boys up in a tree pelted me with hazelnuts. Janet, being smaller, was not allowed to wander far from mum. The Londoners rapidly made themselves at home and soon had the kettle boiling, but it took us a lot longer. As it got darker, fires were lit and groups of people gathered round them, baking potatoes in the embers and drinking endless cups of tea. Grandma and Auntie Lizzie shared our hut for a few days then caught the bus home to look after dad who was still working.

The next day we all went to the hop fields and the procedure was explained to newcomers. The hop fields, filled with row upon row of hop bines climbing up strings which criss-crossed at the top, reached to an enormous height. Huge canvas and pole hop bins stood between each row, ready to be filled. They could be lifted on to a stantion like a hospital stretcher. Men came down each alleyway, pulling down the bines from their string supports with a hook, and left them on the ground for the pickers. Aunt Lizzie was expert and quickly demonstrated the method to mum and me. You grab a bine, rip it off the string then, holding it near the top nearest your finger and thumb, deftly strip the hops off into the bin. I can't remember being any help, but reckon I could still strip a bine with the best of them! I have mum's Hop Pickers' Account Book which records all hops picked each day for six weeks. Mum was picker number seventy-three. As soon as someone had a fair amount of hops in their bins, they gave a shout and a man came along and measured them with his basket. Each basketful was recorded in the account book.

On our first day we picked nine baskets and were paid 8d per basket. We got progressively better with practice and by 18th September reached an all-time high of twenty-nine baskets! On our account book cover is a note scribbled in grandma's handwriting stating that the No. 78 bus ran from Hook Green to Tunbridge Wells every hour, so I imagine many of the women caught the bus into the nearest village to shop for food, rather than rely on the weekly van delivery.

Every morning the women queued at the farmhouse back door to buy the day's milk, each clutching a jug or whatever container they could lay their hands on. I'll never forget the morning I was entrusted with this task: I joined the queue with mum's one and only jug which was duly filled by the farmer's wife. I don't know how it happened, but somehow I dropped the lot and I had to go back to mum and confess. Not only had we to pay for more milk but our only jug was smashed to smithereens! I was mortified. It was all a long time ago, but I only have to close my eyes to conjure up the sights and smells of that time.

As autumn approached it was very cold in the mornings and took more of an effort to climb out of bed. We started picking at 9 a.m. with a short mid-morning break for tea from vacuum flasks, more tea plus sandwiches at lunchtime, then in the early evening a whistle was blown and everyone made their weary way back to camp. My most overwhelming memory is of the smell of hops; a kind of pungent, earthy smell. Occasionally they allowed us into the oast house where they dried the hops. It was lovely and warm there, which brought out the aroma of hops as they dried. Another abiding memory is of sitting round the camp fires at night; it was the only warmth we had on chilly evenings before creeping into our straw bed.

Once a week a van arrived selling groceries and greengroceries for the pickers to top up with provisions. The man also sold sweets and I've never since managed to find a stick of liquorice with that same strong taste. The sticks were hard and lasted for hours, keeping their taste right down to the last mouthful. Another vivid recollection is of the day mum fainted. I'd been off somewhere and when I returned to our bin mum and my little sister were nowhere to be seen. Always a panicker, I immediately began to cry whereupon some kind soul told me what had happened and that mum had been taken to the oast house to recover and get warm. I ran as fast as I could and found her sitting on a chair in that heady warmth, sipping a cup of tea with my sister on her lap. Reassured that she was ok I stayed with her while she recovered, enjoying the comfort.

There are memory lapses of specific details, but I know that mum became friendly with a lady who lived in Lamberhurst village. Perhaps she was a friend of the farmer's wife, I don't know. However, we were all invited to hers for tea one afternoon. Mr and Mrs Robards had two sons, Anthony and Paul. Anthony was about my age and Paul was about three or four. What stands out about that day was being in a bricks and mortar house after weeks of existing in a tin shed. I was convinced they would at least have a proper loo, but NO! I asked to use the toilet and Anthony was instructed to escort me down the garden path to a wooden shed. There it was, yet again – another plank with a hold in it and a bucket underneath! But at least it didn't smell as bad. The family must have known the farmer because the two boys were regular visitors to the hop fields.

Anthony and I would go off to play, leaving mum picking hops and looking after Janet. One day, dad and grandma arrived by bus for a day's visit. I think, though not absolutely sure, that they took Janet back home with them, leaving me with mum and Aunt Lizzie. The last picking recorded in the book was on 9th October 1950, recording ten baskets. The total picked over the six weeks was 621 baskets, yielding a grand sum of £20 18s 4d. No doubt some of the money went to benefit the family but I clearly remember mum treating herself to a new coat when we got back home. It was a swagger coat made up of black, blue and beige squares, and she more than deserved it. I tell this tale simply because it means so much to me. Thank goodness mum kept the account book and the one and only photograph showing her and me standing at the hop bin … I feel that this once-in-a-lifetime experience paved the way, and maybe went towards making me the person I am today, deeply appreciative of the beauty of the countryside and having a strong affinity with all things rural. My final abiding memory is of returning to Coombe Road Junior School, Brighton, some three weeks after the new term had begun (how was mum

A cart that was built to last.

given permission to keep me away all that time?) only to find my class had already started to learn joined-up writing and I had a lot of catching up to do.

Mrs Barbara Roberts (née Leuty)

O Beer! O Hodgson, Guinness, Allsop, Bass!
Names that should be on every infant's tongue!

C.S. Calverley (1831-1884), *Beer*

D. BR...HER.   *Beech*   (486 B)
_____   Farm.
              Marden   Railway Station.
        We commence picking on_____ **30 AUG 1951**____ at 7.0 a.m.
and shall expect you here to take your_____ *2*_____ Bins.
                                                         Baskets.

        Do not come earlier than the date given below, for we shall not
be ready for you. **YOU SHOULD TRAVEL BY THE TRAIN SHOWN.**

_____

                                              LONDON BRIDGE.
        If you are travelling by rail, please be at_____
Station at_____ *5 30*_____in the morning on_____ **29 AUG 1951**
and enquire for_____train.
        Cheap Fares will be available by special Hop-pickers trains.
        Accommodation is arranged for you on this train and it will bring you to
the Station shown at the top of this card, which is the nearest one to our Farm.
        BRING THIS CARD WITH YOU AND SHOW IT AT THE RAILWAY STATION.

Above: A voucher for the Hop Picking train.

Below: The Milk Train used to bring hoppers down to Kent.

CHAPTER TEN

THE PARRY FAMILY AT REAVES FARM AND
RON AND JUNE SPARROW'S REMINISCENCES

We were regular hoppers at Plumford Farm. Plumford lies south of Faversham close to the hamlet of Painters Forstal. For many years it was run by the Elsworthy family. As far as I can make out my gran, Emily Parry, went hopping there from the early 1930s. In fact she went every year until the farm stopped hop production in 1968. She was on the farm when World War Two broke out, as it started in September and that's when the hop-picking season starts. My gran had eight children, all of whom, at one time or another, went with her taking their own families with them, including her son George (my father), and his wife Nellie (my mother). We all travelled to Faversham by lorry, usually Pilcher's coal lorry. We set off from my gran's home in West Street, Gillingham. Weeks of preparation went into this exercise as we had to gather together all the stuff we needed to take down with us. The journey was always memorable and from the 1950s when I was a lad I loved to travel down there; it meant we were going down on our holidays. Although many families went from the Medway towns such as the Perrins, Hendersons and Heards, most families came down from the East End. Most were large families and in hindsight it is difficult to determine where everyone slept. My gran would get her letter from Plumford during July each year, asking her to confirm how many huts she would need. We usually needed three.

Each hut would be allocated a row of hops to be picked per section and only people from your hut would work on that row. Gran would usually visit the hop gardens between Whitsun and September to wallpaper or distemper paint the interior of the huts we would be using and also to visit her sister-in-law, Ada Godden, who spent each spring and summer time on the farm training the hops up the strings.

Once on the farm, the various members of our family would set things up for the time we were to stay there. The farmer would deliver bundles of faggots (small branches used for firewood) by horse and cart each day so we had enough to cook on. Straw for our bedding was delivered by the same means. We cooked on an open fire and all our rubbish was thrown into open 45-gallon drums, which were collected by the same wagon. If it rained we had shared cook houses to use so we could still cook our meals and stay dry.

Mrs Sparrow's father, Jimmy England, to the right of the picture.

The Sparrow family in 1930.

The Sparrow family in 1954.

This page and overleaf: The Sparrow family at Hawkhurst.

These consisted of six post structures with corrugated tin roofs. Our original huts were made of wood planking, tarred on the outside to preserve them, but later they used breeze blocks and cement screed to build us new huts which were much less draughty. All lighting was provided by paraffin lamps.

Food was very basic. For a start we didn't have time to spend on fancy meals but also it was a long walk to the village so we bought only what we could carry. For lunch we mainly made (the night before) cheese or corned-beef sandwiches with stewed tea in vacuum flasks.

Our beds consisted of large hop sacks (pokes) stuffed with straw and laid out on a trestle supplied by the farmer. Children of all ages in our family would have one hut to sleep in, with the boys at one end and the girls at the other, cheesy feet and all. We children would be required to pick hops in the morning and then run any errands for the grown-ups before we were allowed to go off and play.

Errands would mean going to the village shop to collect anything from matches to another tin can of paraffin. Considering it was about one mile to the nearest village, carrying paraffin was not a popular chore, especially coming back again we had to negotiate a very steep track called Stony Alley. The cans had tin handles which hurt your hands when you'd been carrying a while, so we had to keep putting them down and lifting them up again.

Once picking started the farmer would employ some men from the labour force to drive tractors, and they also pulled bines and carried them to the pins for the pickers to strip. My father was chosen to do these jobs on several occasions over the years.

During the day, various 'visitors' came and went at the farm. The Co-op milkman was one and we could buy milk from him for the day. There weren't any fridges so we had no way of keeping it cold and milk was mainly bought day to day. The Co-op baker's van was also a regular caller and daily supplied bread for our sandwiches. Then there was the district nurse who had a sense of humour and would call out 'Bring out your dead!' when she came to attend to any cuts or rashes. On Sundays the vicar came to run Sunday school and usually all the children attended.

We picked our hops into wicker baskets, each of which held a bushel's worth. My gran always made sure they were sound and she'd repair any broken ones, as if they had holes in them the hops fell on the ground and you didn't get paid for them, even though you'd picked them. Wooden stools were made by the family for us to sit on while picking off the hops and we wore rubber aprons to protect our clothing and keep them clean.

It took the workers about five weeks to pull and pick all the hop gardens, which stretched right up the valley for miles. Our family worked for five-and-a-half days each week, but at the weekends all the aunts, uncles, cousins and friends came down to visit us and we'd enjoy the break.

Hops were collected twice a day after being measured by the measure. They were taken by horse and cart to the oast house up on White Hill to be roasted, dried and bagged. Unfortunately this area has been turned into new housing and the oast has disappeared. My family loved their beer so, despite it being a mile away from the farm, they made several visits each week to the 'local', the Alma at Painters Forstal. It is still a very good pub and has hardly changed, even in the twenty-first century.

On Saturday afternoons we children were each given a shilling by our parents and we would walk into the town of Faversham to spend it. Usually we went to Woolworth's. Sometimes we walked via Water Lane to Ospringe, which had a stream running beside it. This disappeared forever when the M2 motorway was built.

Machine picking arrived at Plumford in the early 1960s but our farmer still kept us hand pickers. The machine shed was opposite our huts. Although the huts were demolished many years ago, some remains can still be found, including a double toilet hut. This hut had a long plank with two holes in it sited over a soil pit. The farmer's hand who drove the horse and cart would bring round quick lime and put shovels full down a hole around the back of the hut once a day. There was a wooden flap to cover the hole at the back. It was not very nice if you were sitting in there at the time the lime was being delivered! It was very acrid smelling. Come to think of it, the old boy never smiled but maybe that was because he did have some horrible jobs.

Our water came from stand pipes sited between the rows of huts. We collected ours in large, white, enamel buckets and they weighed a lot when they were full.

We look back and think of our time there as lovely holidays but in reality it was hard work and if it rained, well it was terrible, and of course every morning there was always a very heavy dew so your clothes got soaking wet and bits of hops and caterpillars stuck to you.

I remember one year, probably 1961, when the farmer must have done particularly well, because after the crop was gathered in he arranged a musical evening and treated us all to fish and chips, and we had a right old knees up.

In all the time we went to Plumford I can't remember any trouble between anyone.

In 1998 I took my caravan to Painters Farm Caravan Site and invited all my surviving aunts and cousins along for a walk down to the farm and where the old hop gardens used to be. Many stories were recounted and we had a picture taken of us all standing where

Jim and Emily England with a friend.

Hop pickers arriving at Bodiam by train.

the huts were. It turned out to be a lovely day. I asked Simon Elworthy when my gran first went to Plumford but unfortunately the hop farm records didn't stretch back that far, only to the 1950s, but his mother did recall the Parry name. He did send me his ledger pages for 1955, however, which certainly did mention our name.

Kevin Parry

Ronald and June Sparrow's Reminiscences

I met Ron and June Sparrow at a hop festival weekend held by the Kent and East Sussex Railway at Bodiam in 2006. June's Grandad, Harry Locock, was an Irish tinker and delivered groceries to the Grocery Bar at Calvary Barracks at Hounslow in the First World War and was a regular hopper. Many tinkers went hop picking as it was an easier way to earn a decent amount of money than by the more haphazard method of mending pots and pans. Both Ron and June's families were hop pickers and one year the two families were employed by the same farmer, which is how the two young people came to meet. It was a perfect setting: lovely weather, the beautiful countryside, romantic hop fields with their heady scents and a boy and a girl. After they had met down at the hop field, Ron and June discovered that at the time they only lived a street away from each other in the 1930s, which was a happy coincidence. June was born in 1937 and by that time her family had moved to Greenwich. Before that, her family had lived in Reed Road in Dagenham, Essex. They hopped at a Guinness farm near Bodiam. Aunt Jenne (née Locock) and Uncle Ted Fish usually went hopping at Mereworth, Kent, in the 1950s. Ron's family lived in Hardy Road, Dagenham, but moved after the war and went to live in Peckham. Ron believes his family hopped at Morgan's Farm near Hawkhurst. Their children were Tommy, Jeannie, Ron and Brenda Fish. June's mum was Mary Anne Emily England (née Locock), her dad was James Samuel England and their children were June, Doreen and Jimmy Henry England. 'It's strange the things we recall most, thinking back, but I remember my sister Doreen was always crying', June reminisced sadly. 'Dad was a great individual and had something special about him – he always wore a white, silk scarf and a waistcoat, whatever job he was doing.'

CHAPTER ELEVEN

Every Season was Different

Not long ago I was telling my Australian friend, Colin Christiansen, about hops and the hop-picking saga and he was amazed to hear I was on my fifth book. 'My mind is saying you can only cook and consume a chicken once and that knowledge and experience must run out eventually', he said. But in fact every hopper I meet has a different story to tell and there seems to be no end to the wide variety of fascinating reminiscences hoppers recall.

Burnt Beef and Tatties

When I think back, there was always something different happening. You'd think that there wouldn't be much that could be, as we went to the same place with nearly always the same people every season, but that's not true. Every year we seemed to have something new to remember it by, whether it was when the hut's roof leaked in 1953 and mum and dad got wet through on the top bunk, or the year the farmer's haystacks caught fire and we all had to turn out at four o'clock in the morning with a line of buckets and all in our night clothes to put it out. That was a laugh that was! I just wish I'd been able to afford a camera that time. And there was one memorable year when the cottager where we'd left our Sunday joint (my husband had brought it down from home for Sunday dinner for us) and some parsnips and potatoes to be roasted in her oven for a shilling, and the woman accidentally left it too long in the oven while she was doing some gardening and it got burnt, so we only had a bit of left-over ham from her pantry and some boiled potatoes she gave us to go with our veggies for Sunday dinner. What a disappointment! I love roast tatties and to my little six-year old at the time it was a real tragedy; he wouldn't stop crying and it was a good thing the Lolly Man came round to the huts that afternoon so I could get him a toffee apple.

Nell Hearson

A cookhouse, 1930.

Danger at the Cook House

I can't remember a lot about going hop picking because I was only five at the time this happened, but I do remember the brick-built cook house at Whitbread's Farm down in Goudhurst where we picked each year, and that that was in the middle of the field where the hoppers' huts were that all the pickers used. My little brother, Brian, was only three and we all got a nasty fright when he managed to fall into the cook-house fire. He got a really bad burn all down one side and mum had to take him to hospital to get it seen to. You can still see the scar from the burn on the side of his chest. My sister-in-law was called Bertha and while we were down there hopping, she started going out with a good-looking gypsy from the gypsy camp in the field next to ours. His name was Nelson. The gypsies usually kept separate from the other hoppers at the other side of the field as they had a different way of life to us. Bertha and Nelson courted for quite a while and even got engaged but it never came to anything and they didn't get married. We often heard of romances started down in the hop fields, like Bertha's, and some of them got married in the village church near their hop farm. Quite a few babies were born thanks to going hopping and we always knew about it because they were born about June or July. My mother-in-law and her mum were both pint-sized women and we all went down hopping together. They were full of life and the two of them used to walk down the lanes to the local pub on a Saturday night. It was their one night off and they looked forward to it all week. First, they'd have their drinks while having a good sing-song in the bar with the others, then they'd walk all the way back to the hop farm along the dark lanes to their huts and they'd have a good sing and dance all the way back from the pub. It really made their week. There were no street lights down along the winding country roads and there were deep ditches beside them so it was easy to fall in. It was our family's holiday and it wasn't until mum found it too difficult with a larger family that we stopped going.

Maureen Townsend

Grandfather was a Lolly Man

My grandfather used to sell sweets down at the hop fields. He had a wood tray with a strap that went round his neck to hold it up and all his sweets were a penny each, whether they were lollipops or sherbet dabs. The children all crowded round him when he arrived with his tray of sweets – and so did all the wasps. He must have had a thousand wasps flying round him, trying to get at the sweet smelling, sticky toffee apples. He used to take the family down in his horse and cart and they piled everything they had to take for the time they were down there up on top of the cart. It took a full day-and-a-half to get there and we'd stop somewhere along the way for the night, wherever we had reached that evening. We couldn't afford to stay in a hotel or anything so we slept where we were on the cart. We went all the way to Wheeler's Farm down in Marden in Kent. It seemed such a long way. We set off at about eight o'clock in the morning on the first day and we were too excited to sleep much that night; then we set off again at six o'clock the next morning. The first thing we did when we arrived was to collect our straw and faggots and make up the bed in our hut. You put down the faggots on the floor, covered them with straw to make a mattress and that's what we slept on at night. The day after arriving we had to be up at seven o'clock ready to have our breakfast and start picking. We had the same hut on the same farm every year. There were no raincoats so we used to put sacking round our shoulders to keep the rain off when the weather was wet; we still were expected to go hop picking even if the weather was bad. When the hand picking was finished and the farmers started using machines some of them let the hoppers who had worked for them use their huts as holiday homes so they still went down every year even though there was no picking for them. My grandfather didn't just sell sweets, he also sold metal mugs which were very popular as they were used down at the pub. The landlord wouldn't let them have glasses as they got broken too often, so they had to supply their own mug if they wanted a pint. The metal mugs didn't break as easily so grandfather sold a lot. The local pub was called The White Hart and the hoppers were regulars when down in Marden. We sometimes had a holiday in Ramsgate which folk thought was very posh in those days. It was a nice family seaside town. But most of our holidays were at the hop farm.

Eileen O'Sullivan

The Healey Family at Southfleet

My dad, Thomas King, was born in Southfleet in 1907. He had one brother, Jim, and four sisters. They lived at Waterworks Cottages, Downs Road, Southfleet, where his father worked. The cottages still stand where they were. His father, my grandfather, fought in the Boer War and in the First World War. He was in the Royal Navy. He was only forty-seven years of age when he died of his war wounds and there is a military headstone at Southfleet Church in the graveyard to commemorate his name. My dad left school when he was thirteen to start work on the farms in Southfleet. We have

Above left: Horse-drawing equipment.

Above right: One of the local animals that hoppers may come across.

Left: Mr and Mrs Healey celebrate their engagement in 1926.

two pictures of dad in the hop fields at Northfleet Green. Dad was about seventeen so it was 1924 when the photos were taken. The third photo is of his brother James, or Jim as he was known. Jim was born in 1899 and worked with horses on the farms all his life. Standing behind Jim is their cousin, Sam Llewendon. Sam was born in 1903. He later became a shepherd working for Harris's the farmers at Broadich near Southfleet. He had his own sheepdogs and he would cycle on his bike to work with the dogs running along beside him on the road. He was also the verger of Southfleet Church for many years and went on visits to Jerusalem a couple of times. He died in 1977. Dad worked for French's at Hook Place Farm and my mum was in service there. That is how they met. They got engaged in 1926 and were married a year later at St Mary's Church in Stone, Kent, in 1927. Stone was where my mother lived with her parents, but of course when someone was 'in service' they 'lived in.' (To be 'in service' meant a person worked for a particular family, usually as a maid or housekeeper for a small wage paid annually, for their food and lodging and usually their uniforms, and this meant that unless there were unforeseen circumstances, they had a job for life – something to be prized when jobs were so hard to come by.) She was mostly at Southfleet and was given a tied cottage (which belonged to the farm or estate for as long as they worked for that employer.) This was at Hook Green Cottages where my eldest sister, Stella, was born in 1928. When dad moved farms they went to live at 'The Barracks', Brakefield Road, Redstreet, Southfleet, where my sister Mavis was born in 1934 and then myself, Marion, in 1936. Later, dad left the farms and went to work at Southfleet Waterworks and so they moved to Banbury Villas; by then I was three years old. My mother did casual field work for the local farmers and this included hop picking at Northfleet Green and at Betsham and Westwood. Stella recalls picking hops as a teenager at Westwood and can remember the Londoners coming down to the farms to pick. One day they all went on strike because they said they wanted more money, but Stella said 'I'm pleased with what I'm getting' and carried on picking. Some of the locals warned her 'I would stop picking if I were you or the Londoners will lynch you!' A lady on a horse came by and they said it was Miss Bartholomew, the farmer's daughter and she said something to the Londoners so the dispute was soon settled. Mavis remembers in the war years the women wouldn't go into the hop fields because of all the air raids, but instead they brought the hop bines up to the bottom of our gardens so the hops could still be picked more safely, and when there was an air raid we could easily all run to our shelters. This farm was Chamber's and they were fruit farmers, but the hop fields in Betsham right through to Westwood adjoined this land. They had their own hop kilns which were still working right through the 1960s.

My twin sisters, Brenda and Barbara who were born in 1944, also remember picking in the hop fields. Like me, they can remember sitting in the hop pokes (bins) while the hops were being thrown in by the pickers who were standing all around us. Mum also worked for Gedney's in the summer months and would go fruit picking with dad at the weekends. They would lift me up onto the ladders to pick any fruit at the top of the spindly (thin) branches because I was the lightest! My brother Stuart was born in 1945. He was a Down's Syndrome baby and mum would put him in the big twin pram with my sisters when they were walking to the fields for the day's work. Prior to the twins being born mum would either walk when they were small with us in the pram or when we got older and had learned to ride a bike we would cycle along the country lanes to the hop fields. It was safe for us as there was hardly any traffic then. My dad died aged fifty-seven in 1964 and his brother Jim died in 1961, aged sixty-one. In the photo of mum holding me as a baby, my sister Mavis is on the left of mum and Stella would be about eight years old.

Marion Healey

Up at the Oast House

In the oast house they had a big net made of sacking; very heavy sacking. You'd go up there to turn the hops over with nothing on your feet. But as you stood there and looked down you could see this great big fire underneath you! But you wouldn't fall through or anything because it was strong sacking. The blokes who used to dry the hops never had any sleep; they stayed awake all day and all night because you had certain times for the hops to come out so they didn't get dried too much. Then they had to be shovelled into big sacks using a scuppet and then a man comes with what they call a presser and then they take a sample away to let you know if the hops were good. In them days you used hops for beer but now they're used mainly for dyes. We had a lot to do in them days. I did a lot of cooking because there was always people staying with us – the soldier's wives and sometimes the London hoppers, so I did a lot of baking and fruit bottling. I worked all the year at Grange Farm where we lived.

Alice Heskitt

The nameplate on Crowdleham oast house.

Above: Crodleham oasts.

Right: Inside a roundel roof
of an oast house.

Above: A vane on an oast house.

Below: Oast houses at Shoreham.

A Sunday spent with the Salvation Army.

First Aid and Hygiene

If you got hurt or anything you used to go to the hoppers' hospital to see the nurse. Some places didn't have a hospital; they had the Salvation Army women instead. Mostly it was castor oil to clear your bowels, or syrup of figs if you were lucky because it tasted nicer. But she wanted you back at the hop fields so she got you back there as fast as she could. If you got a cut they'd give it a good wash in case you'd got anything in it, slap a bit of iodine and a plaster on it.

Anon

Mildew and the Bane of the Red Spider

Bines are susceptible to several different blights and one of the most devastating is caused by a tiny red spider. In 1960 we started growing the new Boadicea, which is a variety of hop strongly resistant to mildew. The red spiders tend to leave it alone as the leaves are more bitter than other hop varieties. But the red spider is a menace and now we import a special spider from Africa which eats them.

Peter Cyster, Rother Valley Brewery

Horrible Home Remedies

Of all the home remedies the castor oil was the worst because you'd have to find somewhere to go if you needed the toilet and if we were in the fields we were often a long way from there so you had to hide behind a bush or something and hope no one saw you. Not that it was any better at the latrines because the smell there was terrible. They had this lime pit and you sat over it on a wood plank to do what you had to do and the lime got up your nose and made it sore inside. One kid fell in and we had to wash the lime off him so got buckets of water to wash him. It burns your skin if you don't wash it all off, poor little lad. We didn't have toilet paper or nothing. We'd tear up newspaper pages into squares, thread them by the corner on a piece of string and hang a bunch of it on a nail on a couple of the poles that were holding up the hessian walls. One for the men and one for the women. We called it 'The Ritz'.

Anon

Unmistakable roundels of an oast house.

Right: A fire in a roundel.

Below: Makeshift living quarters for the stoker

THE LAST CHAPTER

Hopping songs abound, many using the same tune but with different words. A song which must have a never-ending list of verses is *Hopping Down in Kent*. Here is one version I've been given (other versions are in some of my other books about hopping):

HOPPING DOWN IN KENT

Now hopping's just beginning
We've got our time to do,
We've only come down hopping
To earn a quid or two.

Chorus: With a Tee-I-O, Tee-I-O, Tee-I-ee-I-O

Now on early Monday morning
The measurer he'll come round,
Saying 'Have your hops all ready
And pick them off the ground.'

Chorus: With a Tee-I-O, Tee-I-O, Tee-I-ee-I-O

On early Tuesday morning
The bookie says with cheer,
'I'll pay you all your winnings
If you stand a pint of beer.'

Chorus: With a Tee-I-O, Tee-I-O, Tee-I-ee-I-O

Now he's taken all our money,
So what are we to do?
We only came down hopping
To earn a bob or two.

Chorus: With a Tee-I-O, Tee-I-O, Tee-I-ee-I-O

They all say hopping's lousy
I do believe it's true.
Since I've been down hopping
I've had a chat or two.

Chorus: With a Tee-I-O, Tee-I-O, Tee-I-ee-I-O

Early Saturday morning
It is our washing day,
We'll boil 'em in our hopping pot
And we hangs them o'er the ground.

Chorus: With a Tee-I-O, Tee-I-O, Tee-I-ee-I-O

The author's grandmother used a mangle
similar to this one.

Now hopping is all over,
the money's nearly spent,
But, thanks to my old china,
I've saved enough for rent.

Chorus: With a Tee-I-O, Tee-I-O, Tee-I-ee-I-O

My arms are scratched all over,
My hands have gone all brown,
But don't you let the hopping
Never, ever get you down.

Chorus: With a Tee-I-O, Tee-I-O, Tee-I-ee-I-O

Nell Hearson and other contributors

A bagging machine at the
Museum of Kent Life,
Cobtree.

Pumpkins and pumpkin fun at Shoreham.

Despite the decline of hop growing in England, bines are still readily available at some of the country museums such as the Museum of Kent Life at Cobtree, The Hop Farm Country Park at Paddock Wood, some countryside shops such as Castle Farm's Hop Shop at Shoreham in Kent, and a few private farmers who grow hops to order for breweries such as Shepheard Neame at Faversham, as well as many private breweries throughout Kent and Sussex who grow their own.

As well as the memories, the recipes live on. Here are a few to wet your appetite:

STUFFED HOP LEAVES
(SIMILAR TO STUFFED VINE LEAVES)

4 large hop leaves, washed, and centre ribs removed
1 large onion, chopped and peeled
40g butter
1 tablespoon double cream
1 egg yolk
Salt and ground black pepper
Pinch of ground coriander
45g cooked rice

Method:

Soak hop leaves in boiling water to soften. Discard water. Pat leaves dry. Lightly brush with olive oil on both sides. Fry chopped onions in butter until soft. Quickly mix in rice, egg yolk, herbs, cream and coriander. Place spoonful of mixture into the centre of each leaf and wrap into parcel. Fasten with a toothpick. Cook for three minutes in remainder of butter. Carefully turn and cook on other side for four minutes. Delicious served with horseradish sauce.

KENTISH HOP BEER

225g hops
450g demurer sugar
Knob of peeled, bruised, fresh root ginger
Yeast
1 gallon water

Method:

Boil hops, sugar and ginger in water for an hour, replenishing water as it evaporates. Strain. When cool add yeast. Stand for twenty-four hours. Syphon off liquid but don't disturb yeast. Bottle, cork and wire the tops. Ready to drink in two days (for a different flavour add 25g cracked maize during boiling stage).

A hope poke (*left*) and vats (*right*) at Larkins Brewery, Northiam. (Iain Ross-Macleod)

HOPS AS A SEDATIVE AND SLEEP INDUCER

When hopping, mothers found their toddlers to be fretting because of long hours in the hop fields, it was regular practice to put them into the sacking hop bins among the cones. Shaped like a cradle with the sides too high for a small child to climb out, they quickly settled in among the picked flowers and fell asleep as hops are soporific. I learned from experience not to put a hop bine in the back of a closed car and drive a long distance as I'd find myself dropping off to sleep at the wheel. Hops are from the same family as tobacco and contain two parts methyl to three parts butane to two parts oil. They have been known by the Chinese for years as a calming sleep inducer and even today it is possible to buy hop pillows as an aid to getting a good night's sleep. Use mature, freshly dried leaves as these are far more potent than when young or have been left dry too long. Mix with dried lavender to make a fragrant herbal pillow. The two herbs combined will help you to slumber soundly without the need for sleeping tablets.

HOP TEA

Use four or five hop flowers and scald with boiling water to make an infusion. Hop tea should not be given to children under three years of age, nor to pregnant women as it is reputed to contain elements similar to female oestrogen.

HOPS AS A STOMACH SOOTHER

Hops are a botanic relative to marijuana and, combined with peppermint, are an ancient cure for upset stomachs. The leaves can be rolled and smoked as a cigarette and taken this way will not cause intoxication. Hops are known as an anti-spasmodic used by the Romans after a heavy night of drinking and were found to be beneficial in soothing the digestive tract. They were keenly recommended by Thomas Culpepper, a well-known herbalist of olden times.

EMERGENCY FIRST AID

In an emergency, a handful of crushed hop flowers can be used on cuts and scrapes until a first-aid box can be found as it is a mild antiseptic. They may sting slightly if used on open wounds.

HORSES, PONIES, CATTLE AND DONKEYS

If you own any of these animals then hops are an invaluable aid to feeding and keeping your animal healthy and in show condition with minimum effort and maximum effect. However, be wary!

When I was a child my mother ran a riding school and we had an arrangement with our local Tetley's brewery to deliver a regular supply of what were known as 'maltings' or malt culms to our stables. Maltings are the mast from which beer is brewed, being the residue of sprouted barley plus the hops used for the purpose in the malting chamber. They have a strong, heady smell of alcohol, having been fermented in vats. The first time we used them, being ignorant of the effect of fermented hops and barley on horses, we tipped the whole delivery into a large feed trough the size of a bath, and let the animals help themselves. The results were rather alarming. The next morning our horses were definitely drunk on their feet, to the extent that we dare not take them out for their regular exercise as they could hardly control one foot in front of the other but still insisted on stumbling round the paddock, somewhat cross-legged and just out of our apprehensive grasp. Believe me, a drunken horse is a sight to see! After that, we only gave out measured amounts and kept the situation under control. The maltings, or culms, are beneficial to the animals and help to keep them healthy. These animals quickly develop a taste for maltings and look forward to it. Choose light, crumbly culms and ensure you are sold fresh supplies, rather than the darker, sometimes dirty type which could well be the sweepings from a brewery floor. To get the best results, maltings should be mixed with other feed such as a couple of handfuls of bran plus one small handful of bone meal to a large bucket of maltings. Add water to dampen it heavily as otherwise the bran and bonemeal, being dusty, will get up the animal's nostrils and cause distress. Mix well. Feed two or three handfuls per evening to a fully grown horse and one large handful to young stock. This will help to keep their coats in good order with a healthy sheen. Brewers grains are another by-product of brewing, being the refuse of malt and can be used as a useful pick-me-up for a jaded animal. They are available wet or dry. It is best to order a small amount of both grains and maltings as both tend to turn sour if they are stored too long. For owners who like to know percentages of the feed they are giving their animals: malt culms have an approximate equivalent of 43.4 per cent starch and one to three ratio of protein. The above amounts are for working animals, not for animals which live mainly sedentary lives. They are a good supplement to feed in winter.

BREWERIES AND BINES

Hops are included in the medical list of safe herbal substances to use, which is as well for many discerning beer drinkers! On a drive through Kent and Sussex you may come across some of the many small, private breweries offering their own brands of beer such as Larkins at Chiddingstone, Rother Valley at Tenterden and Harveys at Lewes. Their beverages are often sold at country fairs, festivals and village fetes and are well worth sampling.

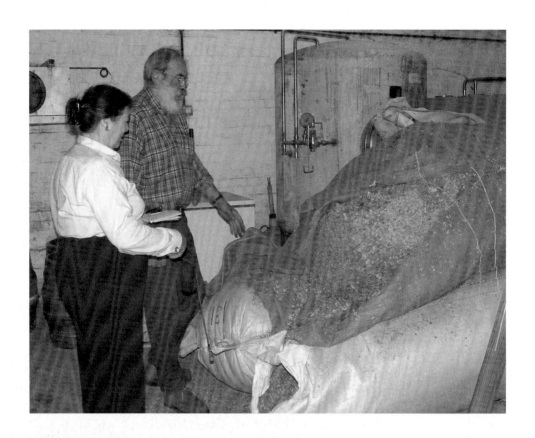

Above: The author and Mr Larkin discussing hops. (Iain Ross-Macleod)

Left: A bagging machine at Larkins, Northiam. (Iain Ross-Macleod)

A converted oast house. (Iain Ross-Macleod)

A hop bine is both useful and decorative. Bines are popular as kitchen or hall decorations and, provided they are allowed to dry slowly and naturally, will last a year in situ. Once the cones are dried, however, it is best to avoid knocking them as the weaker ones will disintegrate. A whole bine can be festooned along a wall as a garland, or small branches hung around individual pictures for ornamental effect.

It is a well-known saying that all good things come to an end and, sadly, this is the end of my book. I hope to be putting out more hopping tales on a talking book soon, so I'm still looking for stories from the hop gardens. If you live in London, Kent, Sussex, Essex or Hertfordshire remember me when you next need a speaker for your club, school or society. All talks are accompanied by photos as a PowerPoint presentation. I give talks on many subjects and look forward to meeting you.

Hilary

Mr Larkin (*above*), a close-up of the tiles on the hop kiln at Little Fowles (*below*), Mr Larkin's hop shed (*opposite above*) and hop-picking machine plate at his brewery (*opposite below*). (Iain Ross-Macleod)

Other local titles published by The History Press

Voices of Kent and East Sussex Hop Pickers
HILARY HEFFERNAN

Right up to the late 1950s, the annual hop-picking season provided a welcome escape for thousands of families who lived and worked in the poorer parts of London, who would migrate every year to the hop gardens of Kent and Sussex to pick the harvest. The photographs and reminiscences in this book tell a fascinating story; of hardship, adventures, mishaps, misfortune and laughter experienced during hardworking holidays among the bines.

978-0-7524-3240-3

Old Kent Inns
DONALD STUART

The old inns of Kent have had a rich history since the first pilgrims came to Canterbury. They have been used as a hide-out by smugglers, highwaymen and deserters, seen the Civil Wars, the era of Napoleon and the bombing raids of the Second World War. Containing more than 90 photographs, this fascinating book tells of spy-holes and smugglers, of cock-fighting, ghosts, buried treasure, murders and hangings, phantom ships, hidden tunnels, hiding places and bodies bricked into walls.

978 07524 3959 4

East End Neighbourhoods
BRIAN GIRLING

The River Thames, with its docks, wharves and associated industries, has been a source of livelihood for generations of East Enders living in the historic riverside neighbourhoods of the former Metropolitan Boroughs of Stepney, Poplar and adjacent areas. From images of the maritime stores of old nautical Limehouse and the silk-weaving houses in Bethnal Green at the turn of the twentieth century, to views of the prefabs in Poplar after the Second World War, this selection recalls how life was lived in the tightly packed streets of the East End.

978 07524 3519 0

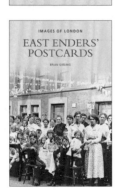

East Enders' Postcards
BRIAN GIRLING

This selection of old views from the capital city's East End combines popular sights with everyday scenes; from aerial views of Tower Bridge and London Docks to vistas of terraced houses, shops and businesses, as well as trams, streets and buildings which have long since disappeared. Images of King George V's Silver Jubilee celebrations in 1935, along with markets at Brick Lane, reflect the traditional community spirit for which this area of London is renowned.

978 0 7524 2494 1

If you are interested in purchasing other books published by The History Press, or in case you have difficulty finding any History Press books in your local bookshop, you can also place orders directly through our website
www.thehistorypress.co.uk